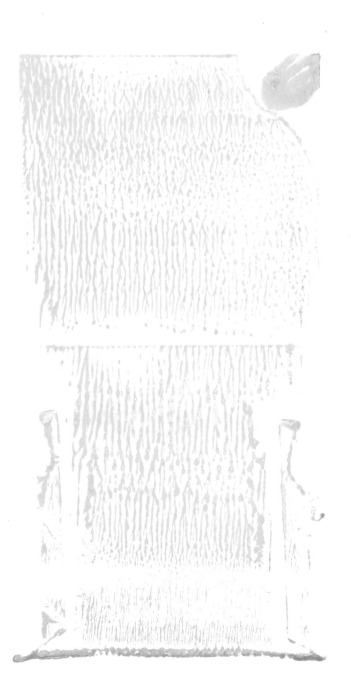

Life History of a Fossil

Life History of a Fossil

An Introduction to Taphonomy and Paleoecology

PAT SHIPMAN

Harvard University Press
Cambridge, Massachusetts, and London, England   1981

Copyright © 1981 by the President and Fellows of Harvard College
All rights reserved
Printed in the United States of America

Library of Congress Cataloging in Publication Data

Shipman, Pat, 1949–
   Life history of a fossil.

   Bibliography: p.
   Includes index.
   1. Paleontology.   2. Vertebrates, Fossil.
3. Paleoecology.   I. Title.
QE721.S54   566   80-29053
ISBN 0-674-53085-3

To my parents,
with gratitude for their steadfast faith in my abilities

Preface

This book focuses on vertebrate taphonomy and paleoecology simply because it is vertebrates that I know best. For the same reason, many of the examples are drawn from the literature on the African sites that are important in human and primate evolution. Because of this emphasis, I have probably overlooked some excellent books or papers, and I invite readers to call such works to my attention.

This work is intended primarily for advanced undergraduate courses in anthropology, paleontology, and evolutionary biology, but graduate students and professionals in those fields should also find it a useful introductory guide. I have tried to provide a clear introduction to the relevant fields of study while avoiding oversimplification of the pragmatic and theoretical difficulties of such work.

I owe much to others in this field. Vertebrate taphonomists have established an atmosphere of open and intimate communication that we hope will not vanish as the field grows. Diane Gifford has characterized this feeling well: "Like other primates who have found themselves in the wide open, uncharted spaces, we tended to huddle together for mutual protection and have consistently found friendly interaction with other members of our own little band to be the best strategy for individual survival." In no particular order, I thank Diane Gifford, Karen Davis, Kay Behrensmeyer, Andrew Hill, Tom Gray, Dorothy Dechant-Boaz, Gary Haynes, Rick Potts, Bill Bishop, Wendy Bosler, Peter Andrews, Judy Van Couvering, John Van Couvering, Richard Klein, Peter Dodson, Meave and Richard Leakey, and many others for friendly interaction. For that, and much more, I am especially grateful to my husband and colleague, Alan Walker.

Bill Bennett and Bill Patrick, my editors, have been most helpful in their roles of coach and cheerleader. I thank Mary Carrington, Kate Francis, and Ok Chon Allison, who typed and retyped the manuscript without complaint. Elise Le Compte and Linda Gardner cheerfully helped me with permissions, the bibliography, and myriad other tasks. David Bichell is responsible for many of the original drawings. The lines from "Ash Wednesday," in *Collected Poems 1909–1962* by T. S. Eliot, copyright 1936 by Harcourt Brace Jovanovich, Inc.; copyright © 1963, 1964 by T. S. Eliot, are reprinted by permission of the publishers, Harcourt Brace Jovanovich, Inc., and Faber and Faber Ltd.

Contents

1 Introduction   *1*
2 Why Do Bones and Teeth Become Fossils?   *17*
3 Geologic Setting and Sedimentary Environments   *45*
4 Spatial Distribution of Fossils in Sediments   *65*
5 Tracing the Taphonomic History of an Assemblage   *99*
6 Faunal Analysis   *123*
7 Postdeposition and Postfossilization Distortion   *171*
8 Some Unanswered Questions   *191*
   Glossary   *199*
   References   *207*
   Index   *219*

Under a juniper-tree the bones sang, scattered and shining
We are glad to be scattered, we did little good to each other,
Under a tree in the cool of the day, with the blessing of sand,
Forgetting themselves and each other, united
In the quiet of the desert.

       T. S. Eliot, *Ash Wednesday*

1 Introduction

What is a fossil? In its broadest definition, a fossil is any trace, impression, or remains of a once-living organism. Thus fossils include the faintest imprint of an early jellyfish or an ancient fern, a cast in rock of a shell, the footprints of a striding dinosaur, the teeth and bones of our own ancestors, and a myriad of other things (Figure 1.1). However, the most common vertebrate fossils are bones and teeth.

Fossil bones and teeth differ substantially from their nonfossilized counterparts. During the process of fossilization these skeletal elements, usually without their accompanying soft tissues, are preserved from further decay. The agents of preservation include mud, sand, and other sediments, lava, rotting vegetation in bogs or swamps, and in unusual circumstances, climatic factors such as extreme cold or dryness. Over the long expanses of geologic time, the organic material in the bones and teeth is slowly impregnated with minerals present in the sediments (see Chapter 7). Therefore, burial in sediments is a first, necessary step in fossilization.

In very special circumstances, animals or plants may be preserved without becoming stone, as in the case of the frozen mammoths (Stewart, 1977), the "bog people" preserved in peat in northern Europe (Glob, 1954), or mummified bodies from Egypt and elsewhere. However, these special cases are rare and have

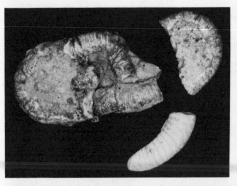

Figure 1.1 Fossils from Fort Ternan. *Top left*, a fossilized millipede; *top right*, two fossil gastropods, *Cesartum* sp; *bottom*, a nearly complete skeleton of a small carnivore, *Genetta* sp. At left are the hindlimbs and parts of the pelvis; in the center are the skull and vertebral column; at the right are the forelimbs and scapulae. (Reproduced by permission of the National Museums of Kenya.)

been preserved by different circumstances from those responsible for lithified fossils. Unless otherwise specified, in this book the word *fossil* refers to the bones and teeth of vertebrates preserved in stone.

Hill (1975:18) elegantly pointed up the single most important feature of fossils. *"Fossil animals are dead.* This fact, though obvious, has significant consequences for the interpretation of fossil assemblages. The mode of death affects many of the other features of bias; it may be responsible for disarticulation and the loss of soft parts, for their concentration and damage, and their burial in sediment" (italics added).

That fossil animals are dead is of tremendous importance, because death brings with it changes in many of the factors that are of interest to paleoecologists. Some of these changes that accompany death can be seen by comparing Figures 1.2 and 1.3. Most fossil animals no longer possess soft tissues like muscles, flesh, and brain; their bones are no longer articulated, and some of their bones are broken or destroyed. Their bones and teeth have been mineralized. Fossil animals do not live in social groups; they have no home range or preferred habitat; and they do not move, feed, play, learn, reproduce, fight, or engage in any other behaviors. Their bones are not associated with those of the animals they interacted with in life. In short, through death most evidence of the interesting information about animals—what they look like, what they eat, how they move, where they live, and so on—is lost. Only through indirect evidence and painstaking study can any information about their habits and lifestyle be reconstructed; this study is called paleoecology.

The ultimate aim of paleoecology is to observe, in all its complexities, the interactions, habits, and lifestyles of a community that no longer exists. Paleoecologists attempt to reconstruct whole communities of interdependent plants and animals and to unravel the web of relationships among the members of those communities. It is not enough to study a single species or a complex of animal species living in the same area, for even preliminary investigations show the influences of geomorphology, climate, and vegetation on a living ecosystem. Ideally, paleoecology should achieve a level of resolution as close as possible to that of studies of modern communities, which in themselves are difficult because of the tremendous complexity of living communities. But studying the ecology of ancient communities is even more difficult, because the only data are those of the fossil record. Therefore it is important to recognize the biases inherent in that record.

The major difficulty in reconstructing the past by analyzing a fossil assemblage is deducing how that assemblage, or collection of

Figure 1.2 A herd of elephants in Tsavo National Park, Kenya.

Figure 1.3 A baby elephant in Tsavo National Park, Kenya, that has been killed by the lioness in the background.

skeletal remains, was derived from the animal communities that once lived in the area. The paleoecologist wants a detailed picture of what plants and animals lived there and how they interacted with each other and the environment. But the fossil record provides an obscure, imperfect glimpse of a few of the original ecosystem's many elements, because not all animals and very few plants are preserved. The geologic setting also has a profound influence on a community's chances of preservation. For these reasons, a paleoecological reconstruction must take into account the effects of all the various preservation factors.

In nearly all cases it is incorrect to assume that a fossil assemblage is a living animal community frozen in time by the process of fossilization. Yet that was implicitly assumed by many older paleontological studies, which took for granted these ideas:

1) *Life environments are equivalent to death environments.* That is, animals are assumed to die in the environment they lived in.
2) *Abundance in the fossil record reflects abundance in the original communities.* In other words, it is assumed that animals found more commonly as fossils were more common in life.
3) *Species found in the same fossil assemblage reflect sympatry of those species in life.* It is assumed that the animals lived together in the same area.
4) *Absence of a species from the fossil record reflects absence or rarity in the original animal community.* Species not found in the fossil record of a particular area are presumed to have been genuinely rare or absent in that area in life.

It is only in the later part of the twentieth century that paleontologists have begun to be aware that these widespread assumptions are often incorrect.

Traditionally, paleontologists have studied two aspects of fossil materials: anatomy and taxonomy. Comparative anatomical studies of similar living and extinct species can do much to reveal the adaptations and habits of the extinct species. In turn, taxonomic studies, showing the relationships among fossil species, are often based on anatomical studies. By comparing related species anatomically, the paleontologist can discover clues to the species' probable positions in the evolutionary history of their group.

In 1940 J. A. Efremov, a Russian paleontologist, proposed a new branch of paleontology, to be called taphonomy. The word is derived from Greek roots: *taphos*, meaning burial, and *nomos*, meaning law. The concerns of this new discipline were described by Efremov: "The chief problem of this branch of science [taphonomy] is the study of the transition (in all its details) of animal remains from the biosphere into the lithosphere, *i.e.*, the study of a process in . . . which the organisms pass out of the different parts of the biosphere and, being fossilized, become part of the lithosphere. The passage from the biosphere into the lithosphere occurs as a result of many interlaced geological and biological phenomena" (1940:85). The major foci of taphonomy are the events that intervene between death and fossilization and the effects of those events on the retrieval of information about the past.

Efremov focused on four major problems of paleontology that can be clarified by taphonomic studies. First, in older strata the state of preservation of fossils often is poorer and the numbers of individuals less, because the chances of destruction increase over time and because sometimes there have been fluctuations in the sizes of the faunas sampled. Second, although a complex of terrestrial mammals found together in one place is usually called a "fauna," a term meant to imply that the animals were associated in life, such a complex may have come together accidentally, either at death (in which case the assemblage is a *thanatocoenose*) or after death (in which case it is a *taphocoenose*). Third, the sudden appearance in the fossil record of a new fauna without clearly identifiable ancestors in the preceding strata may or may not reflect a genuinely new group of species; its sudden appearance may be an artifact of preservation. Fourth, the likelihood of preservation of different body parts or species is so different in different sedimentary environments that complexes of species derived from a single, synchronic, and sympatric community may be taken as different faunas because the original community has been sampled so differently.

These problems, Efremov asserted, may be solved only through the detailed study of localities where terrestrial vertebrates have been fossilized (paleotaphonomy) and through comparison of the characteristics of fossil assemblages with those produced by the processes of death, distribution, concentration, deposition, and preservation of the remains of modern animal communities (neotaphonomy).

Both taphonomy and paleoecology are concerned with reconstructing the past as it pertains to a particular assemblage or fossiliferous environment. Because taphonomy provides the detailed knowledge of events that have biased the assemblage, taphonomic studies must precede the broader reconstructions of paleoecology. For example, it would be insufficient to reconstruct a paleoenvironment as open country simply because large, savannah-dwelling species are present and smaller, forest-dwelling forms are absent; taphonomic studies might show that there has been preferential destruction of all small bones and individuals. Or, when an assemblage is composed of water-sorted and water-transported bones from several different environments, only detailed taphonomic studies can sort out the species into their appropriate communities, each of which may require its own paleoecological reconstruction. Thus it is important to determine whether or not an assemblage has been transported from the place where the animals died and to differentiate between a living community with a network of interdependencies, on the one hand, and circumstantial associations of skeletal elements of animals from different communities of origin, on the other.

Despite the difficulties inherent in reconstructing the past, paleoecology is not a hopeless task. Behrensmeyer, Western, and Dechant-Boaz (1979) recently studied the processes of death and decay in different modern environments in Amboseli National Park, Kenya. They were able to show that the modern bone assemblage closely resembled the living animal community, especially if only species weighing 20 kilograms or more were considered (see Table 1.1). Not only were these larger species present in the assemblage, but their relative abundance in different habitats generally reflected their habitat preferences in life (Figure 1.4). Judging from the Amboseli data, it is possible to conclude that animals sometimes *do* die where they live.

Unless certain cautions are observed, however, it is unrealistic to assume that fossils died and were found where they lived. First, it is essential to have clear evidence that the assemblage has not been transported from its original location by water or other

Table 1.1 Species list for Amboseli National Park, Kenya.

≥15 kg	
* *Loxodonta africana* (elephant) * *Diceros bicornis* (rhinoceros) * *Hippopotamus amphibius* (hippopotamus) * *Giraffa camelopardalis* (giraffe) * *Taurotragus oryx* (eland) * *Syncerus caffer* (buffalo) * *Oryx gazella callotis* (fringe-eared oryx) * *Equus burchelli* (Burchell's zebra) * *Connochaetes taurinus albojubatus* (white-bearded wildebeest) * *Alcelaphus buselaphus cokii* (Coke's hartebeest or kongoni) * *Kobus ellipsiprymnus* (waterbuck) * *Tragelaphus imberbis* (lesser kudu) * *Aepyceros melampus* (impala) * *Gazella granti* (Grant's gazelle) * *Tragelaphus scriptus* (bushbuck)	*Redunca redunca* (Bohor's reedbuck) * *Litocranius walleri* (gerenuk) * *Gazella thomsoni* (Thomson's gazelle) * *Phacochoerus aethiopicus* (warthog) * *Panthera leo* (lion) * *Crocuta crocuta* (spotted hyena) * *Hyaena hyaena* (striped hyena) * *Panthera pardus* (leopard) * *Acinonyx jubatus* (cheetah) * *Felis caracal* (caracal) *Orycteropus afer* (aardvark) * *Papio cynocephalus* (yellow baboon) * *Homo sapiens* (man) * *Bos taurus* (domestic cow) * *Equus asinus* (donkey) * *Ovis aries* (domestic sheep) * *Capra hircus* (domestic goat)

≥1 kg, <15 kg	
Raphicercus campestris (steenbuck) *Rhynchotragus kirki* (dik dik) *Felis serval* (serval) *Proteles cristata* (aardwolf) *Mellivora capensis* (ratel) * *Canis adustus* (golden jackal) *Canis mesomelas* (black-backed jackal) *Viverra civetta* (African civet) *Felis libyca* (wild cat) * *Otocyon megalotis* (bat-eared fox) *Genetta genetta* (small-spotted genet)	* *Atilax paludinosus* (marsh mongoose) *Ichneumia albicauda* (white-tailed mongoose) *Helogale parvula* (dwarf mongoose) *Ictonyx striatus* (zorilla) * *Cercopithecus aethiops* (vervet) * *Galago senegalensis* (bush baby) * *Hystrix cristata* (porcupine) * *Pedetes capensis* (spring hare) * *Lepus capensis* (African hare) * *Canis familiaris* (domestic dog)

Source: Behrensmeyer, Western, and Dechant-Boaz (1979).
* Indicates skeletal remains found in the bone sample.

forces. There is as yet no evidence showing that transported assemblages preserve the species' proportions in life even when the bones have not been moved out of the animals' general habitat. Second, only medium- to large-sized species are likely to be found in their life environments and in proportions reflecting those in life, even if the bones have not been transported. Finally, only the general proportions or relative abundance of fossil species will re-

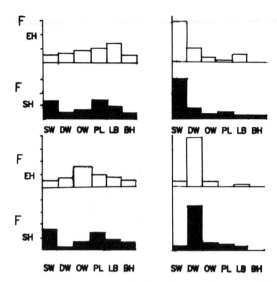

Figure 1.4 Comparisons between the distribution of expected (F_{EH}) and observed (F_{SH}) carcass frequencies across six major habitats in the central Amboseli Basin. Expected carcass frequencies are based on distribution for living animals and annual turnover rates. Observed carcass frequencies are based on the minimum number of individuals found as carcasses in the surface bone sample. *Top left*, zebra (N = 315); *top right*, buffalo (N = 53); *bottom left*, wildebeest (N = 452); *bottom right*, impala (N = 44). SW = swamp; DW = dense woodland; OW = open woodland; PL = plains; LB = lakebed; BH = bush. (From Behrensmeyer, Western, and Dechant-Boaz, 1979; reprinted by permission of the authors and *Paleobiology*.)

flect their frequencies in life. Their absolute numbers as fossils will be far lower than in life, and the number or proportion of any single species may be distorted. Despite these restrictions, the Amboseli study is important because it demonstrates that it is possible to reconstruct paleoenvironments from information about relative abundance of species, given the appropriate conditions.

Historical Sciences

Paleoecology and taphonomy are historical sciences, concerned with retrodiction, tracing past events, rather than prediction of future events. The purpose of such sciences is to find out, as exactly as possible, what *did* happen, to retrieve from the available data—

in this case, the fossil record—information about the past that may not be immediately apparent. It is not a goal of paleoecology and taphonomy to predict the sequence of events that will take a living community from the present, with its wealth of available information, to future fossilization, in which much of that information will be lost or obscured.

Pragmatically, there are major differences between retrodiction and prediction. Although it is often possible to arrive at a retrodictive explanation that makes sense of events that have already occurred, such explanations are rarely adequate as predictions, for two reasons. First, the position of taphonomy and paleoecology within the empirical sciences must be considered. In Medawar's (1974) model, physics, chemistry, biology, and ecology can be arranged hierarchically (as in this list), with each science being a subset of the one that precedes it. The scope of each successive science is more restricted than that of the one preceding it, and the flow of ideas, laws, and generalizations is one way, from physics through to ecology. Thus all laws of physics are valid in and applicable to each of the other sciences, but the laws of ecology are not likely to be applicable to or valid in the preceding sciences. As the scope of the science decreases, the complexity increases. For example, one can readily predict the behavior of a steel beam subjected to a given force in a given direction, but it is extremely difficult to predict when (and why) a predator will select a particular prey animal. Thus biological and ecological rules—on which paleoecology and taphonomy are based—have taken the form of probability statements. When Medawar's model is modified to include taphonomy and paleoecology (as in Figure 1.5), it becomes apparent that it is difficult to formulate predictions in these sciences simply because their scope is so restricted.

A second reason for the difficulty of formulating predictions (or their retrodictive equivalent) is that our present techniques and ability to reconstruct past events and conditions are insufficient to yield the necessary amount of detail.

It is important to realize, however, that modern ecological studies also suffer from severe limitations. Ecologists trying to answer evolutionary questions concerning the adaptive significance of physical or behavioral traits are handicapped by their lack of time

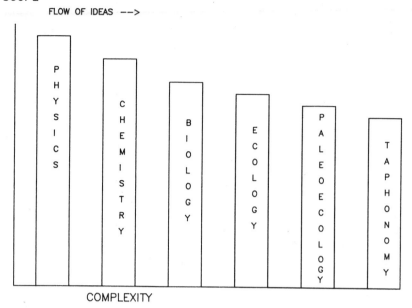

Figure 1.5 A modified diagrammatic model of the sciences, based on Medawar (1974).

depth. Even small mammals that reproduce two or three times yearly can very rarely be observed for the hundreds of generations necessary to document the adaptive advantage or disadvantage of a particular trait. Ecologists working without regard to the fossil record can document abundantly the adaptations and interactions of species, but they cannot view these in the context of the evolutionary history of the species; paleoecologists, on the other hand, have a great deal of data on evolutionary history but little of the complementary data on physical and behavioral adaptations. These two areas of science must interact so that the biological detail can be integrated with the long-term evidence of its impact.

Simpson (1970) pointed out that all historical sciences rely upon the fundamental uniformitarian principle that the present is the key to the past. This principle might be called the first law of taphonomy and paleoecology. However much species and communities evolve and change, the physical principles dictating the survival, transport, and destruction of bones do not change. However, uniformitarianism was initially developed as a geologic principle, and unlike biological phenomena, geologic phenomena operate according to immutable physical and mechanical laws. For this reason, the uniformitarian emphasis now is on understanding how processes produce effects rather than on simply observing the effects and assuming that past processes were like those of the present.

In using uniformitarianism to explain events that have already occurred, taphonomists and paleoecologists must follow three logical steps (Simpson, 1970): (1) obtaining and ordering the historical data; (2) determining what processes are operant in the present and how they produce effects; and (3) confronting the historical record with the knowledge of present processes. The first two steps parallel the subdivisions of taphonomy: the study of the fossil record (paleotaphonomy) and the study of contemporary processes of death, decay, and deposition (neotaphonomy). The third step is the interpretation of the fossil record.

If the first law of taphonomy is: "The present is the key to the past," the second law must be: the occurrence of a past event can be deduced only by demonstrating that its effects differed from those of other, similar events. In other words, only the distinctive aspects of an event can be considered diagnostic. It is not sufficient to show, for example, that a fossil assemblage is not different from a group of bones collected by a porcupine *unless* it can also be demonstrated that its characteristics do differ from those of assemblages collected by other agents, such as a river or a hyena. Modern and fossil assemblages must be analyzed in the same way, with full awareness that the differences between them, caused by both chemical composition and differential preservation, need not obscure the basic attributes and characteristics of the assemblages.

Evidence versus Theory

In one sense, the taphonomic history of a bone assemblage can be taken as a story of information loss (Figure 1.6). At one end is a whole, living animal with habits, abilities, and preferences that are of great interest. At the other end is perhaps a few fossilized bits, often broken, transported, and falsely associated with bones of individuals the animal never saw in life. It is a dismal transformation. However, it is increasingly possible to retrieve lost information through taphonomic studies, or at least to delineate areas in which the fossil assemblage may provide a distorted picture and require cautious interpretation. What taphonomic studies provide is a new source of evidence about the past.

In paleoecology and taphonomy, as in all sciences, it is critical to distinguish clearly between the evidence itself and the theories based on that evidence. It is almost always possible to think of several possible theories to account for a given set of data. If the theory and the evidence are not distinguished clearly, it is difficult to evaluate the probable correctness of the different theories.

In science it is impossible to know that a given theory is true, but a theory can be proved false by gathering evidence that is incompatible with it. Or the data may support a particular theory or interpretation, but the possibility always remains open that new evidence will render it inadequate. Therefore the highest level of confidence one can achieve is that a particular theory is the best (most probable) explanation of the presently available evidence.

The importance of distinguishing between evidence and theory can be demonstrated by an example. It has long been noticed that some fossil assemblages contain bones of only small animals. This is the evidence. Such concentrations of microvertebrates have been explained by a variety of theories. One is that the original bone assemblage has been winnowed by water. It is well established that in general the size and density of a sedimentary particle determines its potential for transport by water. Therefore it is reasonable to assume that bones of animals within a particular size range will behave similarly in a river current and might be transported, and deposited, as a group. Evidence to support such an interpretation might come from a study of the sedimentary environment of the site. Questions that might be asked include: Are the sedimentary particles of the right size and density to have been transported and deposited with the bones? Is there geologic evidence of a stream or river channel?

Another theory that might explain a deposit of microvertebrates is that they are the remains of individuals that drowned in their burrows. Evidence for this theory might be geologic evidence of burrows. Also the fossils ought to be complete or nearly com-

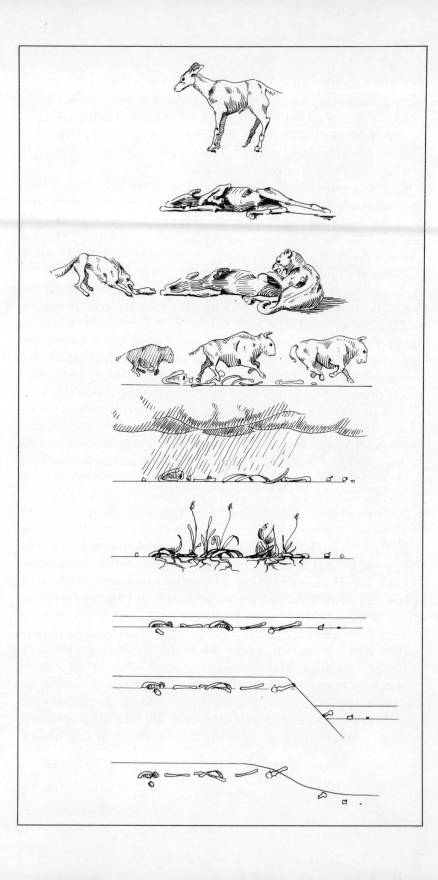

plete skeletons, and at least some of the bones should be articulated, because the remains have been protected from postmortem disturbance.

Still another theory might be that the ecosystem sampled in the fossil deposit contained few or no large animals, like many modern forest environments that are rich in small species and poor in large ones (Bourlière, 1963). Evidence to support this theory might come from taxonomic and anatomical studies of the species represented. It would be important to know whether the animals showed adaptations similar to those of modern forest-dwelling species and whether the fossil species were related to modern forest dwellers. Analysis of fossil pollens and of the geological setting might also prove useful in deciding whether or not the original habitat was forest.

Another possible explanation is that the assemblage represented a population of a single species that was killed by epidemic disease or some other catastrophe. In this case, all individuals represented would be of one species; there ought to be many articulated skeletons; and the population structure ought to reflect catastrophic mortality (see Chapter 2).

Mellett (1974) suggested yet another interpretation of certain microvertebrate assemblages. He collected and examined modern carnivore scat (feces) and owl pellets (regurgitated material). In both he found considerable quantities of bones of small animals. The similarity of such bones to some fossil microvertebrate assemblages is striking (Figure 1.7). Mellett suggests "that most or all microvertebrate fossil accumulations first passed into or through the digestive tracts of carnivores (mostly mammalian, but including predacious fish, reptiles, and birds), and deposited as fecal droppings (scat) in or near stream, lake, or other basins, where they were subsequently covered by sediment" (1974:340). Various sorts

Figure 1.6 (*Opposite*) Information loss in taphonomy. The sequence shows the transition of a whole animal to fossilized fragments of bone. *Top to bottom*, a living herbivore; the carcass of the herbivore; disarticulation and destruction of the carcass by predators; further destruction by trampling; cracking and splitting of bones through weathering; invasion by plant roots; burial in sediments; fossilization; displacement and breakage of fossils by faulting; exposure by subsequent erosion. (Drawing by Dave Bichell.)

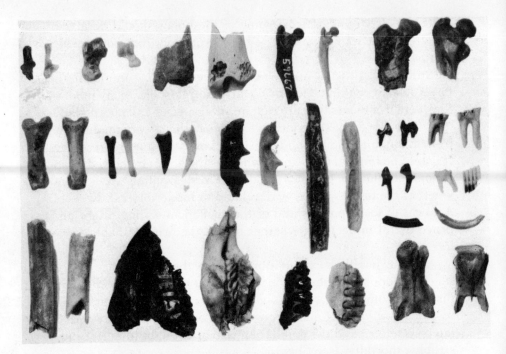

Figure 1.7 Comparisons of microvertebrate fossils (left in each pair) with modern microvertebrate bones from carnivore scat (right in each pair). (From Mellett, 1974; reprinted by permission of the author and publisher. Copyright 1974 by the American Association for the Advancement of Science.)

of evidence could support this interpretation. Geologic evidence might show that the paleoenvironment included a stream, lake, or other basin, but this would not be conclusive. Microscopic study of the condition of the bones would reveal whether they had been exposed to digestive fluids. If the scat itself is preserved as coprolites from which small bones can be excavated, the agent of accumulation is indisputable.

It should be clear that a single set of evidence can be interpreted in several different ways. More important, if only the interpretation is given, it is easy to assume the existence of certain types of supporting evidence that do not in fact exist. Failure to differentiate between evidence and theory makes it difficult for any one but the original researcher to evaluate the interpretation.

2 Why Do Bones and Teeth Become Fossils?

Every year hundreds of thousands of bones enter the pool of possible fossils, but only a few will actually become fossils. All others will succumb to one or another of the many processes of destruction that usually overtake bones, whether in the postmortem period between death and burial, in the postdepositional period between burial and fossilization, or in the postfossilization period. Why are particular specimens preserved and not others? For example, why have so many jaws of australopithecines been preserved at East Turkana, Kenya, and so few scapulae? This chapter will consider the variety of factors that in part determine which bones will be preserved and which will not. Many of these factors influence not only which bones survive the postmortem period but also which are preserved in accumulations of sufficient size to be called fossil assemblages. These factors include both agents of damage or breakage and agents of concentration or collection of bones.

Death

Initially, bones and teeth are parts of a living animal, their shape dictated by the species' ancestry and its habits. Teeth and bones grow, wear, strengthen, or break in response to the events of an individual animal's life, within the limits of heredity. When the animal dies, its bones and teeth become vulnerable to the interplay of the forces of preservation and destruction.

The mode of death itself determines in part what happens to the bones. Individuals in a population usually die off one by one from various causes; this is called attritional mortality. If preserved as fossils in one location, the remains of such a population will demonstrate a true sampling of death rates in different age classes over time. A population that has died through attritional mortality will show high death rates in the most vulnerable age groups (the very old and the very young) and lower rates in the less vulnerable age groups. On the other hand, when a catastrophe such as flood, drought, disease epidemic, or volcanic eruption strikes down many or all members of a local population (catastrophic mortality), the preserved remains of the population will reflect the total age structure of that group in life. The assemblage will probably show large numbers of young individuals, and decreasing numbers in each successive age class.

The cause of death does not necessarily indicate whether the pattern of mortality is attritional or catastrophic. Disease may kill an entire population or only the weaker individuals. Similarly, a sudden flood may cause mass drowning, or it may kill a few weak individuals during an ordinary river crossing.

The most common causes of death of wild animals are: predation, disease, senility, accident, and starvation and dehydration.

PREDATION

Predation is so dramatic that it receives an undue amount of attention (for example, see Ardrey, 1961, 1967, 1976). Predators often choose as targets individuals that differ in some way from the rest of the prey population (Mueller, 1975; Salt, 1967). Predators do weed out the old, infirm, or unfit, but these are individuals that would probably die soon of some other cause in the absence of predators (but see Rudnai, 1973; Mech, 1966; Makacha and Schaller, 1969).

The tendency to inflate the importance of predation is fostered by studies of large predators. For example, Kruuk (1972) found that hyenas kill or scavenge 1.5 to 5.5 kilograms of meat a day for each adult hyena; thus a large pack may have to kill frequently to get enough food. Schaller (1972) reported that a mother cheetah with cubs killed a Thomson's gazelle almost every day, or approximately 337 per year. Mech (1970) reported that a wolf pack of 15 to 17 individuals killed one moose every 3 to 3.7 days during five winters of observation, which would be 98 to 122 moose per year if moose were available as prey during the summers. These numbers are well documented and impressive, but from the point of view of the prey population, the predation rates are very low. For example, the prey killed by Kruuk's entire hyena population in a year (1966–1967) constituted an average of less than 10 percent of the total mortality in each of the prey species. Total mortality for those species ranged from 4 to 16 percent in the years 1959–1966, which means that the predation by hyenas accounted for the deaths of a maximum of 1 to 2 percent of the total prey species population per year. Even for a prey species that lives in large social groups, years —perhaps an individual's entire life—may pass without a single incident of predation within the group.

Since predators usually consume part or all of the carcass, individuals dying of predation have a less than average chance of being preserved as fossils. But the amount of the carcass that is consumed varies, even with a single type of predator, prey, and environment. For example, the amount of a deer carcass consumed by wolves varies with such factors as the rate of successful hunts, the size of the wolf pack, and the availability and size of the prey (Haynes, in press). These generalizations are likely to hold true for other species as well. Many fossils do bear marks of predation or scavenging; therefore at least some bones of animals killed by predators survive to become fossils.

DISEASE

Disease may be a significant cause of death in wild animal populations during seasons or years when food resources are scarce and animals are in a weakened physical condition. When different populations are isolated from one another, the spread of disease is slow and sporadic. But epidemic disease may endanger several populations that are concentrated in a small area by drought (Shipman, 1975) or some other ecological condition. Evidence of disease, whether it results in attritional or catastrophic death, is nearly impossible to find in the fossil record, except in those few cases in which the disease produces distinctive bony changes or in which soft tissues are preserved. Death by disease has little effect on the chances of preservation of the skeletal remains.

SENILITY

Death by senility is probably rare in wild animal populations, since aged individuals are likely to die of disease, predation, or accident rather than old age per se. It is most likely to occur in very large, relatively solitary species, like bison or rhinos. Their isolation protects them to some extent from contracting disease, and their size makes them nearly invulnerable to predators after the first few years of life. Death by senility may be postulated but not proven in cases of fossilized individuals with exceptionally worn teeth; however, it is impossible to tell whether the immediate cause of death was age, predation, or disease. Because individuals of some species lose minerals from their bones with age, those that die of senility may have a lesser chance of preservation. But this probability may be more than counterbalanced by their greater adult size, because there is a strong bias toward preservation of bones of large individuals.

ACCIDENT

Accidents that may cause death include injuries from fighting, falling, being trampled, drowning, or from mudslides or volcanic eruptions. Several natural traps or sinkholes containing large accumulations of bones are known from the fossil record; these include some of the South African australopithecine sites that were once limestone caves with small, intermittently closed shafts to the surface. Animals also die and are preserved in tar pits, as at La Brea, California. The La Brea assemblage is notable for the unusually high proportion of carnivores, which were apparently attracted by the dead bodies of prey species and then were caught in the tar themselves.

The type of accidental death influences the chances of preservation. Accidents that result in immediate burial often produce fossils, as at La Brea or the human settlement preserved by volcanic ash at Pompeii. Drowning or falling into a sinkhole may protect the carcasses from scavenging and weathering, thus preserving them until they are buried by sediment. Other types of accidents, however, especially those that disable the animal, render the victims unusually vulnerable to predation and scavenging, increasing the probability that the skeletal elements will be partly or entirely destroyed before preservation can occur.

STARVATION

Even more than dehydration, starvation is the main cause of death during a drought. Even when all available forage has been consumed, animals refuse to leave a shrinking water source. For this reason carcasses are numerous near waterways and are often preserved when water levels later rise (Shipman, 1975). Mass deaths from an ecological change such as drought produce a surfeit of food for predators, increasing the chances that some bones will not be harmed by scavengers and will be preserved as fossils.

Postmortem Destruction and Transportation

Whether or not a bone is preserved after an animal's death, and in what condition, depends in part on the bone's *preservation potential*—a balance between the tendencies toward preservation and toward destruction that is characteristic of that skeletal element. Of course, the events to which the bone is subjected also influence its chances of preservation, but it is possible to predict the overall likelihood of preservation for each body part. Several sorts of evidence suggest that preservation potential is primarily a function of a bone's size, shape, composition, and other physical characteristics rather than what happens to it after death—its taphonomic history. The way each type of bone fragments, weathers, and breaks when chewed is predictable, suggesting that its structural strengths and weaknesses determine in part its reaction under stress.

Evans (1973) summarized the extensive literature on the various experiments that have been performed to determine the mechanical properties of different bones when subjected to tension, compression, shear, and torsion. In many instances, the different mechanical properties of two types of bones—say, a femur and a fibula—can be directly related to aspects of their microscopic structure, such as the density of canals for blood vessels or the proportion of newly deposited and poorly mineralized bone. Further support comes from Shipman's (1977) observation of a statistically significant relationship between the shapes of broken bone fragments and the skeletal elements from which they are derived. Similarly Tappen (1976) demonstrated that cracks and splits produced by natural weathering reveal the bone's basic structural patterns (see also Lakes and Saha, 1979). In addition, how the bone was broken and its condition at the time influence the shapes of the resulting fragments (Sadek-Kooros, 1972; Bonnichsen, 1979); see Chapter 7 on breakage.

BONE SIZE

The size of a bone is most conveniently measured as its volume, which has a significant influence on its attractiveness to predators, its hydraulic behavior, and its vulnerability to breakage. Because different aspects of size (length, breadth, and thickness, as well as volume) are important in different instances, I will use the more general term *size* here.

At one extreme, very small bones are especially vulnerable to consumption by predators or to collection by small scavengers like harvester ants (Shipman and Walker, 1980). If the bones are concentrated in one place by such agents and then redeposited in or near a sedimentary environment, their chances of preservation are increased. Many consumed bones are protected from further destruction because they are surrounded by undigested fur or fecal matter (Korth, 1979; Dodson and Wexlar, 1979; Mellett, 1974). However, consumed bones may be wholly dissolved by gastric juices or redeposited far from a sedimentary environment and thus escape preservation. Very small bones are likely to be either abundant or very scarce at a particular fossil site. Small bones are also probably more vulnerable to disintegration through trampling and weathering (Behrensmeyer, Western, and Dechant-Boaz, 1979; Korth, 1979).

Small- to medium-sized bones are the most accessible to predators that chew or gnaw bones; smaller bones are usually more fragile and likely to become fragmented (Binford and Bertram, 1977).

The Hyaenidae, which have specialized as bone eaters (Kruuk, 1972; Bearder, 1977; Dorst and Dandelot, 1970; Ewer, 1973), are remarkable for their bone-crushing adaptations and for their ability to digest bones. Other carnivorous species, such as vultures, lions, wolves, foxes, wild dogs, leopards, ratels, jackals, skunks, and wolverines, also consume bones. So do such diverse noncarnivores as sheep, giraffe, cattle, porcupines, rodents, pigs, deer, and other ungulates, for reasons ranging from nutritional deficiencies to the need to wear down ever-growing teeth. The differential survival of skeletal elements subjected to chewing has been linked by several authors (Brain, 1970, 1976; Behrensmeyer, 1975; Hill, 1975; Binford and Bertram, 1977; Shipman and Phillips-Conroy, 1977) to both the size and the composition of the bones.

In addition to being broken, consumed bones may also be transported: as a result of a struggle between predator and prey, as with crocodiles (Behrensmeyer, 1973); in the digestive tract of predators (Mellett, 1974; Sutcliffe, 1970; Dodson and Wexlar, 1979); or, occasionally, when the predator stores a carcass in water (Kruuk, 1972:119). Predators that are fairly large relative to their prey may also carry bones or entire carcasses in their mouths, apparently to chew them in isolation (Kruuk, 1972: Haynes, personal communication; Van Lawick-Goodall and Van Lawick, 1970; Shipman and Phillips-Conroy, 1977).

At the other extreme of the size range are bones too big to be attractive to the available predators. Skeletal elements that are usually very attractive because of their marrow content or associated soft tissues, such as ribs, may remain intact if they are very large (Figure 2.1).

What happens to a bone in water (its hydraulic behavior) is also influenced by its size, especially at the extremes of the size range. Very small bones are unlikely to remain in a sedimentary environment long enough to be buried and preserved, and very large bones are unlikely to be transported into a sedimentary environment at normal current velocities (see "Hydraulic Behavior" below).

BONE COMPOSITION

Bone composition is often incorrectly termed "bone density." By definition, density is the ratio of weight to volume. But bony tissue is primarily hydroxyapatite, $Ca_{10}(PO_4)_6(OH)_2$, which has a constant

Figure 2.1 Skeletal remains of a juvenile elephant in Tsavo National Park, Kenya. Note the lack of predator damage to even the more fragile elements such as the scapula and ribs.

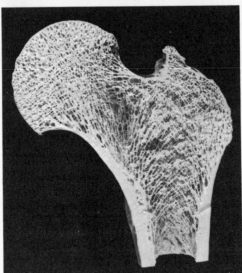

Figure 2.2 This coronal section of a proximal human femur shows the compact bone of the walls of the shaft in contrast with the cancellous bone of the proximal end.

density of 3.1 to 3.2 (Berry and Mason, 1959). When the density of a whole bone is measured experimentally, the different values are the result of the different amounts of space between the bony tissues. Cancellous, or spongy, bone is generally less dense because it incorporates more spaces (Figure 2.2). In life these spaces and the medullary cavity are filled with hemopoietic tissues, which disintegrate after the animal dies, leaving spaces that may become filled with air, water, or mud, depending on where the bone is lying. Thus, when a water-soaked bone is measured, its density includes the water that has infiltrated the open cells and cavities. If the bone is broken or abraded, some of these spaces may be exposed, and the measured density of the specimen may change, although the true bone density will not.

In predicting the potential for preservation, it is probably more accurate and more useful to consider the ratio of spongy to compact bone (the S/C ratio) in a skeletal element than to consider the density of the entire bone under specific environmental conditions. Each type of skeletal element has a characteristic range of S/C ratios, although there is variation among species, because each species' bones are subjected to different types of stresses.

Composition affects the bone's chances of both survival and burial. It has been shown that carnivores prefer spongy to compact bones (Hill, 1975, 1980; Haynes, 1980). For example, they devour the proximal ends of primate humeri more often than the distal ends, because there is a higher proportion of spongy bone in the proximal end (Brain, 1970, 1976). Spongy bone is also more friable than compact bone and thus is unlikely to survive undamaged in areas where predators are active.

Klein (1975) maintained that hominids are more likely than carnivores to break into the shafts of long bones to retrieve marrow; carnivores usually smash the articular ends. Haynes (1980) suggests that each of the major North American predator species has a distinctly different approach to eating a whole carcass, and so the remains of their prey include differing proportions of body parts and different marks on those parts that survive.

The interaction of S/C ratio and size explains the well-known high preservation potential of teeth. Teeth, made of dentine and enamel, have a higher density than bony tissue and are generally small compared to other skeletal elements of the same animal. Their small size means that they may be transported into a sedimentary environment fairly easily, and their high density and low S/C ratio help them remain in the environment until burial.

BONE SHAPE

One important measure of the shape of a particle or bone is the ratio of its surface area to volume (SA/V ratio). Although it is difficult to measure the surface area of an irregular object, a reasonable approximation can often be made. Volume is more simply derived by determining the volume of water displaced by the object. For a sphere, both of these parameters can be calculated if the radius of the sphere is known:

$$V = \frac{4}{3}\pi r^3 \quad \text{where } V = \text{volume of a sphere and} \\ r = \text{radius.} \quad (2.1)$$

$$SA = 4\pi r^2 \quad \text{where } r = \text{radius and} \\ SA = \text{surface area of a sphere.} \quad (2.2)$$

A sphere has the smallest SA/V ratio of any solid, calculated as

$$\frac{SA}{V} = \frac{3}{r} \quad (2.3)$$

so all bones have higher SA/V ratios than a sphere. Because the volume of nonspherical objects is more easily measured than the surface area, it may be useful to compare objects by solving for V:

$$V = \frac{(SA)r}{3}. \quad (2.4)$$

Bones with high SA/V ratios are also especially attractive to predators because of the ease with which they can be broken to extract marrow. Such bones also often have high S/C ratios. Because they tend to be thin and flat, they are especially vulnerable to breakage and in fossil assemblages are likely to be fragmented or broken.

Another convenient means of measuring shape is the shape index (SI), derived by the following formula:

$$SI = \frac{\text{maximum length}}{\text{maximum breadth}} \quad (2.5)$$

with maximum length and breadth measured at right angles to each other. Shape index was first published by Hill and Walker (1972) in their analysis of the Miocene assemblage from the fossil site at Bukwa, Uganda. They cautioned that "by referring only to area this index fails to reveal a number of relevant features. For instance, a cubic bone would have the same index as a square flat one, or a long flat fragment of mammal bone as a long thin cylindrical limb bone" (1972:403). Despite this difficulty, *SI* does provide one useful gross measure of shape and has the advantage of being derived from measurements that are easily and often taken. Because shape is such an important factor in a bone's behavior in water, the distribution of *SI*'s in an assemblage can be used to deduce whether sorting by moving water has occurred.

Few fossil assemblages have been analyzed in terms of *SI*. A detailed comparison of the *SI* data from Fort Ternan, Kenya, collected in 1974, and from Bukwa, Uganda, collected in 1970, will demonstrate some of the deductions that can be made on this basis. Both are Miocene sites; Fort Ternan is dated to 14 million years (Evernden and Curtis, 1965; Evernden et al., 1964), and Bukwa to 23 million years (Bishop et al., 1969; Bishop and Chapman, 1970), respectively. The conditions of deposition were very different, however: the Fort Ternan site was a small, boggy area near the junction of forest and savannah, and the Bukwa site was a muddy lake bed that preserved a forest fauna. The two excavating crews used similar techniques.

Figure 2.3 shows the *SI*'s of 1,149 specimens from the 1974 Fort Ternan collection (data from Shipman, 1977; Shipman, Walker, Van Couvering, Hooker, and Miller, in press). Many of the bones are square or nearly square: 40 percent have an *SI* less than or equal to 1.9. There is a steady decrease in the number of bones as *SI* increases, but bones with high *SI*'s—that is, very long, thin bones—are reasonably well represented.

In contrast, the *SI*'s of 178 Bukwa bones, shown in Figure 2.4, show a different distribution (data from Hill and Walker, 1972). Fully 67 percent of the Bukwa bones have an *SI* less than or equal to 1.9. The shaded histogram in *A* shows the contribution of turtle scutes to this peak of nearly square bones. Also, unlike the Fort Ternan bones, the Bukwa bones are discontinuously distributed in the *SI* categories. Faunal differences alone are insufficient to ac-

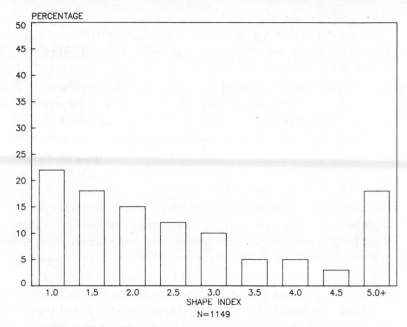

Figure 2.3 Shape indices of 1,149 specimens from the Fort Ternan fossil assemblage.

count for the differences in *SI* distribution at these two sites; *B* shows the Bukwa sample without turtle scutes, which are not present at Fort Ternan. The peak in *SI* categories below 2.0 persists, as does the sharp decline in frequency as *SI* increases. This persistent pattern supports the idea that the Bukwa assemblage was sorted by hydraulic forces according to the shape of the bones, while the Fort Ternan assemblage was not. This interpretation agrees well with the geologic evidence of deposition at these sites.

HYDRAULIC BEHAVIOR

Bones differ in their potential for transport and dispersal in water; this potential is referred to in general as their hydraulic behavior, and, if current action is important, as their hydrodynamic behavior. The differences in hydraulic behavior means that the bones' chances of reaching and remaining in a sedimentary environment also vary. Because burial in sediments is a necessary stage in fos-

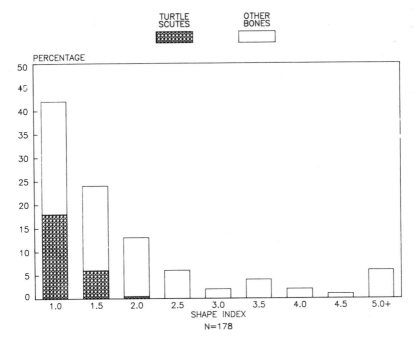
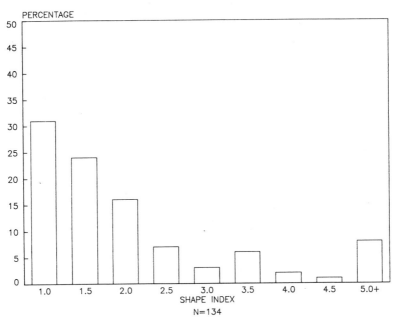

Figure 2.4 *Top*, shape indices of 178 specimens from the Bukwa, Uganda, fossil assemblage. *Bottom*, shape indices of 134 Bukwa specimens. Turtle scutes have been excluded, but the peak at 1.0–2.0 persists.

Bones and Teeth Become Fossils / 29

silization, and because water is often an important factor in the deposition of bones in sediments, the behavior of a bone in water is an especially important influence on its preservation potential.

In general, bone composition and size are the most important factors influencing hydraulic behavior, although one component of shape, the SA/V ratio, is crucial. Bones with high SA/V ratios for their size and composition, such as scapulae and innominates, have a high potential for hydrodynamic dispersal. The higher the SA/V ratio, the more the behavior of a given object or particle deviates from the predicted hydraulic behavior of a sphere of equal volume. A sphere provides the least surface area per unit volume and is therefore more difficult to lift and suspend than a more projecting object. The viscous drag of fluid on a particle obstructing the flow depends on the shape and size of the particle and on its orientation relative to the current.

Figure 2.5, top, shows a scapula in a current moving from the left, which produces an upwardly rotating drag on the scapula. If the current were from the right, the scapula would project less abruptly into the current, the speed of the water passing over the scapula would be less, and the force of the upward drag would be less. However, the current would still produce an upwardly rotating lift, counterbalancing, in part, the pull of gravity that keeps the scapula on the bottom.

Figure 2.5, bottom, shows a sphere in the same current. The upward drag will be less than that on the scapula because of the decrease in SA/V ratio. But what is most important is not the total SA/V ratio but that part of the surface area that is perpendicular to the flow of the current; this is sometimes termed the "shadow area" of the bone.

Shotwell (1955, 1958) was among the first to suggest that the representation of different taxa in a fossil assemblage is often a function of hydrodynamic transport (see Chapter 6). As a first approximation of the rules of hydraulic behavior of bones, Shotwell's method provides some excellent insights. But at the heart of the approach is a tacit assumption that the degree of sorting is a function of the distance over which the bones have been transported. This assumption has been criticized because it does not take into account factors such as current velocity, sediment load, and en-

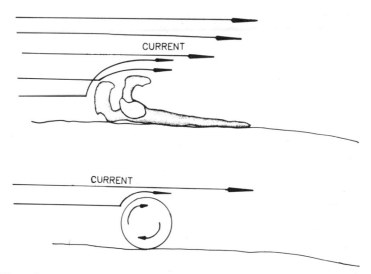

Figure 2.5 Top, a scapula lying in a current. The arrows indicate the upwardly rotating drag on the scapula as the current passes over the bone. *Bottom*, a sphere in the same current is subjected to less upward drag because its *SA/V* ratio is lower than that of the scapula.

vironment of deposition. However, Shotwell's approach started paleontologists thinking in terms of hydrodynamic transport and its consequences for bone assemblages. Those who have built upon his theory with experimental work have emphasized the importance of three factors Shotwell ignored: the size, density, and shape (*SA/V* ratio) of the individual bones.

Voorhies (1969), Behrensmeyer (1975), Boaz and Behrensmeyer (1976), and Korth (1979) have focused on the hydrodynamic transport of bones in artificial streams or flumes. After observing the behavior of bones of medium-sized animals (sheep, coyote) in a flume, Voorhies defined three groups of skeletal elements according to their potential for hydraulic dispersal (Table 2.1). The Voorhies Groups show which skeletal elements are likely to be transported and deposited together by hydraulic forces, and they can be used to deduce the degree of transportation and sorting that has occurred during the formation of a fossil assemblage. For example, an assemblage containing all three Voorhies Groups is probably untransported (autochthonous) and therefore a good subject for

Table 2.1 Potential of different bone types for dispersal by water.

Group I: immediately removed by low-velocity currents; high SA/V ratio, high S/C ratio	Group II: removed gradually by moderate currents; low SA/V ratio, intermediate S/C ratio	Group III: lag deposit, moved only by high-velocity currents; low SA/V ratio, low S/C ratio
Ribs	Femur	Skull
Vertebrae	Tibia	Mandible
Sacrum	Humerus	
Sternum	Metapodia	
Scapula[a]	Pelvis	
Phalanges	Radius	
Ulna	Mandibular ramus[a]	

Source: Voorhies (1969); reprinted by permission of the author and *Contributions to Geology*.
a. These elements are intermediate between the two groups.

paleoecological reconstruction. An assemblage containing only Group III elements is probably a lag deposit of bones left behind once all the smaller animals and more easily transported elements have been winnowed out by the current. A bone's potential for hydraulic dispersal is directly related to its shape; most of the Group II bones are long and roughly cylindrical, with the exception of the pelvis, scapula, and mandibular ramus, all of which, like the long bones, have moderately high SI's.

Behrensmeyer (1973, 1975) undertook to translate these Voorhies Groups into more quantifiable terms by examining the hydraulic behavior of bones in the same way geologists have traditionally examined the behavior of sedimentary particles. The two most important concepts in her approach are settling velocity and critical shear stress (or threshold drag). Settling velocity is the rate at which a particle falls through fluid, commonly measured using a column of clear, standing water. Settling velocity may be used to compare the hydraulic behavior of different particles. The standard formula to predict settling velocity (V_s) of a large particle (comparable to a bone) is the Impact Law, also known as Rubey's Law:

$$V_s = \sqrt{\frac{4}{3} g \left(\frac{\rho_p - \rho_f}{\rho_f}\right) r}$$

where V_s = settling velocity;
g = gravitational constant or 980 cm/sec^2;
ρ_p = particle density (weight/volume);
ρ_f = fluid density; and
r = radius of particle. (2.6)

This formula shows that the settling velocity of a particle is a function of its density and size (radius) and of the density of the fluid, with allowance for the pull of gravity. Unfortunately, this law is formulated to deal with sedimentary particles that are presumed to be spherical—a shape that few bones approximate.

Behrensmeyer estimated the settling velocity of nonspherical particles such as bones by first finding their nominal diameter: the diameter of a sphere of quartz having the same volume as the nonspherical particle. If the volume of the nonspherical particle is known, the nominal diameter can be calculated from the following formula:

$$d_n = \sqrt[3]{\frac{6 \text{ (bone volume)}}{\pi}} \quad \text{where } d_n = \text{nominal diameter}, \quad (2.7)$$

or
$$d_n = \sqrt[3]{1.91 \text{ (bone volume)}}. \quad (2.8)$$

This simple transformation enables one to derive the radius (half the diameter) and thus to estimate the settling velocity of nonspherical particles, although shape and density introduce some inaccuracy.

The viscous drag of fluid on a particle helps to initiate movement by producing lift; it also helps to keep the particle in suspension by slowing its settling. Bones with a high SA/V ratio move at lower velocities and remain suspended at lower velocities than bones with low SA/V ratios, and thus bones with high SA/V ratios are likely to be winnowed out of a sedimentary environment unless the current is slow or the bones are very large.

The S/C ratio also has a major impact on the behavior of bones in water. Figure 2.6 shows what will happen to the settling velocities of bones with a low S/C ratio and bones with a high S/C ratio as their sizes increase. The nominal diameters of these bones approximate the range of those measured for several modern East African mammals (data from Behrensmeyer, 1975). The general trends in this figure are more important than specific values. High density can be taken to indicate a low S/C ratio, and low density to indicate a high S/C ratio. What this figure shows is that size increments have a greater effect on the settling velocity of bones with high S/C ratios than of those with low S/C ratios.

But what use is it to estimate the settling velocity of a bone? Calculating settling velocity is one means of predicting the Voorhies Group of bones for which there are no experimental data. For a bone to remain in suspension, the velocity of the current must be about twelve times the settling velocity for that bone. When the current velocity falls below that point, all skeletal elements with similar settling velocities will settle out of suspension and be deposited. Settling velocity groups can be expected to approximate Voorhies Groups.

Equally important is the concept of critical shear stress or

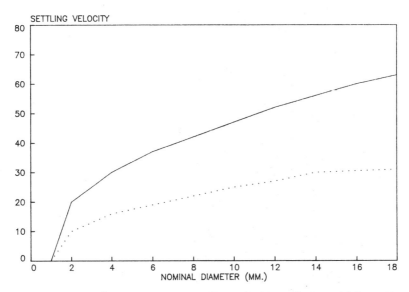

Figure 2.6 With increasing size, settling velocities of bones with low *S/C* ratios increase more slowly than those of bones with high *S/C* ratios. Nominal diameter indicates the size of a quartz sphere that has the same measured volume as the bone.

threshold drag, also borrowed from geologic studies. Critical shear stress is the current velocity at which a given particle at rest on a stream bed will begin to move. Thus it measures the current velocity at which winnowing of a bone assemblage will begin. Behrensmeyer (1975) gave the following formula (after Allen 1970:50) for bones or other particles that are considerably larger than the substrate on which they rest:

$$\tau_{(crit)} = \frac{1}{6}(\rho_p = \rho_f)g\, D_1$$

where $\tau_{(crit)}$ = critical shear stress;
ρ_p = particle density;
ρ_f = fluid density;
g = gravitational constant or 980 cm/sec^2; and
D_1 = diameter of the particle to be moved.   (2.9)

Bones and Teeth Become Fossils / 35

Like settling velocity, critical shear stress is a function of the density and size of the particle and the fluid density, with an allowance for the pull of gravity. Along with formula 2.6, this formula shows that size and density are the most important determinants of a particle's hydraulic behavior.

Experiments by Behrensmeyer and Hanson (Hanson, 1980) suggest that density is much more important than size in transporting bones. In flume experiments an astragalus of a juvenile hippopotamus moved at the same mean velocity as a sheep astragalus when bone density, orientation relative to the current, mode of movement, and hydraulic conditions were the same (Figure 2.7). Although the bones had the same density (defined here as weight in water divided by bulk volume), the volume of the hippo astragalus was twenty-seven times that of the sheep astragalus. It is apparent, then, from this example that density is a more important factor than size in critical shear stress. This relationship is probably less clear with sedimentary particles of the same size and material, because their density is constant. To reiterate, difficulty arises in ap-

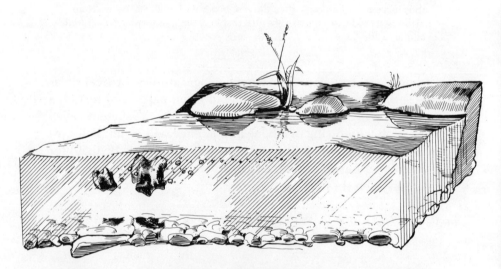

Figure 2.7 A sheep astragalus and a juvenile hippo astragalus will move at the same mean current velocity, because their densities are the same even though the volume of the hippo astragalus is many times greater. (Drawing by Dave Bichell.)

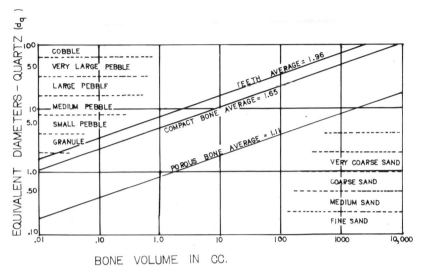

Figure 2.8 A log-log graph showing the diameters of quartz spheres that are hydraulically equivalent to bones of different densities. (From Behrensmeyer, 1975; reprinted by permission of the author and the *Bulletin of the Museum of Comparative Zoology.*)

plying formulas 2.6 and 2.9 to bones because the formulas assume spherical particles.

Figure 2.8 summarizes the relationships among bone volume, bone density, and nominal diameter, and Figure 2.9 shows the relationships among nominal diameter, current velocity, and transport. These figures permit prediction of hydraulic behaviors based on known bone volumes. How accurate are these predictions? Behrensmeyer (1975) tested them experimentally and found that measured settling velocities deviated from the expected by a mean of 18 percent. For some bones the prediction was accurate; for others the deviation was as high as 74 percent. A significant factor in the accuracy of the predicted settling velocities was the SA/V ratio. The more the shape of a bone differs from spherical, the more the observed settling velocity will differ from the predicted one. For example, a large flat bone like a pelvis settles more slowly than predicted by the formula, especially if dropped into the water flat side down.

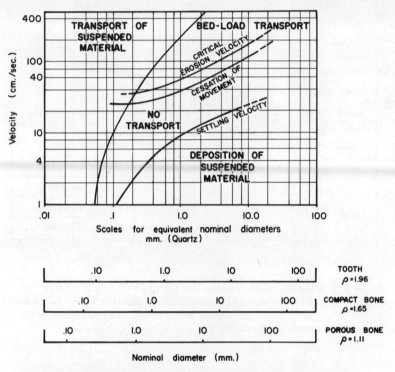

Figure 2.9 Critical shear stress for bones of given densities. Nominal diameters can be estimated from bone density by reading upward from the scale at the bottom to the quartz equivalent on the abscissa. (From Behrensmeyer, 1975; reprinted by permission of the author and the *Bulletin of the Museum of Comparative Zoology*.)

Density is also important in predicting the hydraulic behavior of bones, because it affects both critical shear stress and settling velocity; slight variations in density may change hydraulic behavior significantly. Measured wet densities for modern bones and teeth range from approximately 1.0 to 2.3 (Behrensmeyer, 1975).

What do these rules about the hydrodynamic transport of bones reveal about the taphonomic history of an assemblage? Under most hydraulic conditions, the bones in an assemblage will sort according to their Voorhies Group, which, in the absence of experimental data, can be predicted from the density and volume of the bones.

If all three Voorhies Groups are represented in a given assemblage, this indicates that current velocities at the site of deposition were insufficient to move the smallest and lightest elements. This does not mean that the most mobile elements have not been moved at all, but simply that the velocity at the depositional site was low enough to permit them to settle and be buried. An assemblage containing all three groups provides a maximum of paleoecological information, because there has been little mixing of different animal communities, and few of the easily transported elements have been winnowed out. Behrensmeyer (1975) concluded that normal stream current velocities (20–150 centimeters per second, Leopold et al., 1964) are sufficient to move bones of East African mammals such as sheep, reedbuck, forest hog, topi, and zebra. However, the large bones of the biggest animals, such as hippo, elephant, and rhino, are likely to be moved only at flood velocities.

Dodson (1973) tested the behavior of bones of very small vertebrates—mouse, toad, and frog—in hydraulic conditions. He found that velocities of 6–35 centimeters per second were sufficient to move all the bones, so even the slightest water movement will remove the bones of such animals far from the site of death. He concluded that because these bones are so easily transported, they have little or no use as paleoecological indicators, unless there is good evidence that water played no role in deposition.

Korth (1979) followed up on Dodson's work by quantifying the density and settling velocity of bones of animals varying in size from a mouse to a rabbit. He found that the Voorhies Groups defined on medium-sized animals generally held true for small vertebrates. Therefore, he predicted that assemblages of very small vertebrates transported by currents would show hydrodynamic sorting. Also, small animals are often swallowed whole by predators and the bones transported long distances before redeposition through regurgitation, as with owls, or defecation, as with many mammals (Mellett, 1974). Further, the scats or pellets themselves may float long distances.

Raptor pellets or carnivore scat may preserve the bones of small mammals living within several miles of the point of deposition. The animals represented are likely to have the same daily activity cycle as the predator; owls, for example, rarely catch diurnal species. Harvester ants collect small mammals with a much broader spectrum of habits and ecological niches and are able to forage over much smaller areas (perhaps a quarter of an acre). In contrast, assemblages concentrated by fluvial processes are likely to be nearly useless as indicators of the local habitat.

Korth (1979) documented the representation of various skeletal elements in pellets from several owl species and in coyote scat. The raptor pellets showed similar patterns of representation distinct from that in coyote scat. Comparable work by Shipman and Walker (1980) on bones collected by harvester ants showed a third distinct pattern. The three agents of collection achieved a mean percentage representation (the mean of the number of skeletal elements preserved divided by the number expected in a whole skeleton) between 65 and 78 percent. Korth compiled similar data for hydrodynamically concentrated assemblages, which showed a much lower mean percentage representation (12 to 17 percent). Using Korth's approach, it is possible to distinguish among these various methods of accumulation for small mammal bones. Such distinctions are important in making paleoecological deductions on the basis of small vertebrates.

At Olduvai, M. D. Leakey (1971) found the articulated skeletons of mice and frogs preserved in their burrows. In such circumstances, hydraulic transport is not a real possibility, and there is excellent evidence that the animals have died in the environment in which they lived. But only in such exceptional cases can small mammals be used as evidence of the paleoenvironment.

It is possible to generalize about the hydraulic behavior of bones both by observing how they move in experimental situations and by adapting standard geologic formulas used to calculate such important parameters as settling velocity and critical shear stress. Such generalizations can be helpful in determining whether or not the bones in a fossil assemblage have been transported or sorted by water. This information provides a basis for estimating the usefulness of the assemblage for paleoecological reconstruction. Knowledge of the influence of hydraulic forces on bones can also alert the researcher to possible biases in the assemblage, such as the absence of all small animals and easily transported bones.

Transport may also occur for other reasons. Hominids and perhaps other animals may transport bones for use as tools and weapons (Dart, 1959), as a raw material for building (hominids, Klein, 1969; bowerbirds, MacDonald, 1973) or in conjunction with other activities. (In all these cases the size and composition of the skeletal remains may be important.) Elephants have been reported to transport and scatter remains of their own species, but it is not clear whether they select particular body parts (Dorst and Dandelot, 1970; Douglas-Hamilton and Douglas-Hamilton, 1974). Transportation by itself does not usually damage bones, but it does affect their chances of preservation by changing the environment in which they must survive and be buried.

Postdepositional Events and Fossilization

Of the bones that are not destroyed by postmortem events (see Table 2.2), only a few reach a sedimentary environment suitable for preservation. In order to be fossilized, a bone must not only survive the destructive events that occur, it must also make its way into an environment in which active forces of preservation are at work. Only a very small percentage of all bones survive, become buried, and undergo the slow process of fossilization, which re-

Table 2.2 Effects of postmortem events on bones.

Destructive force	Effects
Predators and scavengers (including hominids)	Consumption, gnawing, breakage
Use of bones as tools	Breakage, wear
Hydraulic transport	Winnowing of assemblage, abrasion
Subaerial transport (rolling, sliding)	Abrasion, breakage
Aeolian transport	Pitting, winnowing of assemblage
Weathering	Cracking, crumbling, exfoliation
Decay by chemicals, roots, insects, soil, water	Disintegration, breakdown of structure

quires special geochemical conditions and an unusual sequence of events.

Even after burial, most bones disintegrate because the soil or sediment condition and groundwater chemistry are not conducive to fossilization (see also Chapter 6). For example, forest animals are not often preserved, because the acid soils of forests cause buried bones to disintegrate. However, fossils are rarely preserved in soils, being more common in fluvial sediments. Yet under the right environmental and geologic conditions, enormous numbers of bones may be preserved. For example, in an area conducive to preservation that supports 1,000 animals, a 10 percent annual mortality rate will provide roughly 10,000 bones yearly to the potential fossil pool. If only 0.1 percent of these are actually fossilized, in 100 years there will be an assemblage of 1,000 fossils. Analysis of the accumulation of bones at Fort Ternan, Kenya (Shipman, 1977) suggests that it represents the untransported and unconcentrated remains of animals that died in a small area conducive to fossilization. Although it is not possible to estimate the time span covered, it is obvious that the assemblage of over 11,000 fossils could have been formed through attritional mortality in a relatively short (in geologic terms) period of time—only a few hundred or thousand years.

Summary

The preservation potential of a bone is the balance between the likelihood of destruction and the likelihood of preservation that is generally characteristic of that element. Preservation potential is primarily a function of the bone's physical attributes: its durability, attractiveness to predators, and hydraulic behavior, each of which is a function of the bone's size, composition, and shape.

The abiotic environment influences the preservation potential of a bone through the pH of the soil or water and the accessibility of sedimentary environments. The biotic environment includes the other species that inhabit the area in which the animal dies, which may consume, gnaw, break, help decay, or transport bones. Which of these interactions occurs depends both on the physical characteristics of the bones and on the species that make up the local animal community.

Finally, the taphonomic history—the catalogue of destructive or preservative forces that actually affect an assemblage—affects a bone's preservation potential. Bones that are buried immediately after the death of the animal are exposed to fewer destructive forces; therefore, their preservation potential is high. Not only are the bones likely to be preserved undamaged, but all types of bones are equally likely to be preserved. An example of such an optimal preservation situation is the fossil assemblage at the La Brea tar pits in California. Assemblages with less favorable taphonomic histories show differential representation of skeletal elements and, among those bones preserved, differential damage. If the mode of death involves rapid burial, as in drowning in a flooding river, this may have a dramatic positive influence on the preservation potential of the bones. At the other extreme, an assemblage may undergo repeated incidents of destruction and transport, such that few bones survive to be buried and fossilized.

3 Geologic Setting and Sedimentary Environments

Geologists recognize three basic types of rocks: sedimentary, igneous, and metamorphic, each type characterized by the process that formed it. It is important to know these types of rock and their formation processes in order to understand their role in preserving fossils.

Sedimentary Rocks

One way in which rocks are formed is by the lithification of loose sedimentary particles—material that has settled out of the water or air. Lithification usually occurs by the binding together of the particles in a process termed cementation. Nearly all fossiliferous beds are composed of sedimentary rocks (also called sediments), because burial in sediments does much to prevent destruction of bones and teeth.

Sedimentary particles are divided into the standard size categories given in Table 3.1. To a great extent, the size of a sedimentary particle determines its behavior in settling out of the medium around it, although particle density and shape are also important (see Chapter 2). There are three categories of derived sedimentary particles or rocks: detrital sediments; chemical sediments; and organic sediments.

Table 3.1 Standard size categories of sedimentary particles.

Particle	Diameter (mm)
Boulder	Greater than 256
Cobble	64–256
Pebble (gravel)	4–64
Grit	2–4
Sand	$1/16 - 2$
Silt	$1/256 - 1/16$
Clay	less than $1/256$

Source: Zumberge and Nelson (1972); reprinted by permission of John Wiley and Sons.

Detrital sediments are composed of particles of other rocks cemented together; for example, sandstone is formed by the cementation of sand-sized grains, usually quartz or silica. Detrital sediments are further categorized by the size of the sedimentary particles. In decreasing order, these are conglomerates, grits, sandstones, siltstones, and shales. Conglomerates are made up of rounded particles from pebble to boulder size; grits are composed of grit-sized particles; sandstones are made up of sand grains; siltstone is compacted silt; and shale is compacted clay. Fossils are commonly found in both sandstone and shale. Breccia is a special type of detrital sediment in which the particles are of the size found in conglomerates but angular in outline rather than rounded, indicating that they have not been transported by water far enough to produce rounding. Breccias may include bones and teeth among their particles, as in the South African australopithecine caves.

Chemical sediments are formed from materials that precipitate out of water: limestone, dolomite, rock gypsum, and rock salt are examples. Both limestone and dolomite are alkaline rocks that foster preservation of bones and teeth. Rock gypsum and rock salt are formed through the evaporation of salt water rather than through precipitation; for this reason, they are referred to as evaporites.

Organic sediments are derived from fragments of plant or invertebrate animal remains that compact into rocks. One example is organic limestone, formed of the remains of coral reefs or shell fragments, both of which are rich in calcite (calcium carbonate).

Another common organic sediment is coal, produced from compacted swamp vegetation. Both of these are excellent environments for the preservation of vertebrate fossils and may contain larger, more complete material than is preserved in other environments. Other types of sediments containing recognizable vertebrate fossils are not usually considered organic, because skeletal remains are only a small proportion of the particles that make up the sediment.

Igneous Rocks

Igneous rocks, formed by the cooling of molten magma, are of great importance in the fossil record for two reasons. First, the subset of igneous rocks known as *pyroclastics*, fragments of rock that have been blown out of a volcano during eruption, often preserve skeletal remains. Pyroclastics range in size from large fragments, known as bombs, to smaller cinders, and finally to fine fragments of volcanic ash. Ashfalls, when consolidated, are referred to as tuffs and may weather into soils that preserve bones, as at Fort Ternan, Kenya. In remarkable instances, tuffs may preserve both footprints and fossils, as at Laetoli, Tanzania (Leakey, 1979; Leakey and Hay, 1979). At Pompeii volcanic ash has preserved the impressions of soft tissues, as well as houses, furnishings, and other inanimate objects. Second, lavas and tuffs are useful for radiometric or paleomagnetic dating, which can give a general idea of the age of the fossils preserved in them or in adjacent strata.

Both radiometric and paleomagnetic dating are widely used to determine the age of rocks older than 100,000 years. Radiometric dating depends on the fact that radioactive atoms in the original environment are captured as the rocks solidify. Such atoms degenerate at a steady rate by giving off particles from their nuclei. Over time, these atoms become either a new isotope of the original element or a new element. The time elapsed since the rock has lithified can be determined by measuring the proportions of the original parent atoms and the new daughter atoms. Of course, if the rocks are reheated to a molten state after their original lithification, the proportions of atoms will reflect the most recent heating and cooling, not the original one. Elements commonly used in radiometric dating are potassium (K) and two isotopes of argon (Ar^{40} and Ar^{39}).

Paleomagnetic dating is based on the fact that as rocks are formed, the magnetic particles in it are aligned by the earth's magnetic polarity. This fact is useful for dating because at various times the earth's polarity has abruptly reversed itself so that magnetic north becomes magnetic south and vice versa. When these reversals last for at least 40,000 years, they can be detected as brief *events* or longer *periods* of normal polarity, as at present, or reversed polarity. The timing and duration of the periods of normal and reversed polarity, punctuated by events, are well documented on a worldwide paleomagnetic time scale. To establish the time span during which any particular stratum was laid down, one samples a sequence of strata from that area and compares the thickness of these strata and their polarity with the paleomagnetic time scale. Detailed knowledge of the stratigraphy is important. Temporary breaks in sedimentation or removal of sediments through erosion alters the apparent duration of the polarity periods as recorded in the rocks and must be corrected for.

Metamorphic Rocks

These complex rocks are produced by the chemical and/or physical alteration of either igneous or sedimentary rocks at high temperatures and pressures. Vertebrate fossils are rarely, if ever, recognizable in metamorphic rocks, even if they were present in the original sedimentary rocks. Therefore, they are not of special interest to paleontologists.

Geologic Environments

In which geologic environments is preservation most likely to occur? Bishop (1963, 1980) has defined what he calls "suitable environments for death" for animals "intending" to become fossils. The following discussion is largely based on Bishop's account. The four major continental settings in which fossilization commonly occurs are:
1) Tectonic areas in which there is rapid and cyclical sedimentation

2) Volcanic fields in which alkaline pyroclastics are given off
3) Continental margins, especially where rivers empty into oceans, forming deltas or lagoons
4) Inland basins and traps that provide temporary or permanent burial in sediments

TECTONIC AREAS

Tectonic activity, or movements of the earth's crust, may induce cyclical sedimentation in two types of environments: fan deposits at the edges of mountainous areas or on the flanks of emerging mountain ranges; and areas that have been thrust downward, such as rift valleys. The former represent tectonic uplift of a source area for sedimentary particles; the latter, the downward movement of an area that receives sediments.

Fan deposits, named for their shape, are commonly formed by the deposition of riverborne sediments as a river flowing down from a mountainous area comes abruptly onto a level or gently sloping plane (Figure 3.1). Vertical uplift, as during the formation of mountain ranges, will cause the raised area to be subjected to

Figure 3.1 A fan deposit is formed by riverborne sediments at the junction of mountainous and nearly flat areas. (Drawing by Dave Bichell.)

more intense erosion than the surrounding low area. As a result of increased exposure, more sedimentary particles are formed, carried downstream, and deposited in a fan deposit. Tectonic activity may also fragment and abrade rocks in the zone of movement, another source of sedimentary particles. In any case, a tectonically unstable area undergoes cycles of increased sedimentation. A depositional area that is lowered through tectonic activity becomes a basin into which rivers drain, carrying sediments. An example is the Rift Valley of eastern Africa, which serves as a sediment trap for an enormous area. Such catchment areas can be excellent environments for preservation of fossils if the sediments are alkaline. Because sedimentation in both types of tectonic environment is closely tied to hydraulic transport, the resulting fossiliferous beds are fluvial, meaning that they are deposited by flowing water.

Many fossiliferous strata, or beds, are of fluvial origin whether or not tectonism has been a factor in their formation. Sedimentary environments associated with flowing water are channel fills, point bar deposits, flood plain (or overbank) deposits, oxbow lake fills, and deltas. Alluvium (a collective term for fluvial sediments) is deposited in rivers that can be classified as one of two types, meandering or braided.

Sinuous, meandering river channels are often found when the downhill slope is not steep. Figure 3.2 is a diagram of a meandering river, showing the areas where sediments are deposited. Over time the river changes course many times, and the broad, flat valley is formed of old, infilled channels. These valleys are called floodplains because when the river floods, the overflow spills over the banks and covers the plain. Bones lying there may be covered with river sediments or transported, depending on the topography and the water velocity. Overbank deposits result when the water velocity is low enough to leave a fine-grained deposit lateral to the channel. These often contain bones and teeth, as in some of the Siwalik sites in Pakistan (Pilbeam et al., 1979). An oxbow lake is formed when a bend in the river is cut off and then slowly fills by the settling of fine sediments. Point bars develop at bends in meandering rivers when the current erodes the outer, concave bank and deposits sands and grits on the inner, convex bank, where the current velocity slows slightly. Coarser material is left as channel lag deposits, and finer material is swept downstream.

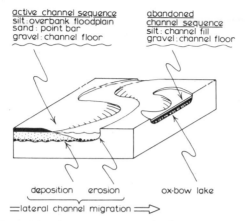

Figure 3.2 Depositional sequences typical of a meandering river. (Reprinted from Richard C. Selley, *Ancient Sedimentary Environments*. Copyright © 1970, 1978 Richard C. Selley. Used by permission of the publishers, Cornell University Press and Chapman and Hall, Ltd.)

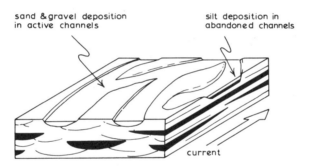

Figure 3.3 Typical deposits laid down by a braided river. (From Richard C. Selley, *Ancient Sedimentary Environments*, copyright © 1970, 1978 Richard C. Selley, reprinted by permission of the publishers, Cornell University Press and Chapman and Hall, Ltd.)

Braided river channels are networks of intersecting channels of low sinuosity (Figure 3.3). They commonly occur today in areas with steeper slopes and more available sediments than are characteristic of meandering rivers. Often such channels are found in arid areas where there is little vegetation to impede erosion. A new channel cuts rapidly into the underlying substrate, but the large amount of sediment quickly fills up that channel, forcing the river to cut yet another channel. Thus a network of channels is formed, crossing and recrossing a whole depositional area. Overbank deposits are relatively rare, although abandoned channels may fill up in the same way as oxbow lakes. In general, deposits left by braided rivers have far less silt and clay than do those of meandering rivers, because of the higher water velocity of the braided rivers.

Table 3.2 summarizes the characteristics of deposits associated with different fluvial environments, which occur both in tectonically unstable areas and in various types of inland basins and traps.

VOLCANIC FIELDS

Airfall tuffs and waterlaid tuffaceous sediments may preserve fossils on floodplains, volcanic slopes, or in small basins or lakes. Alkaline pyroclastics and deposits secondarily derived from them are responsible for the preservation of bones in many sites from the East African Miocene, such as Napak, in Uganda, and Rusinga, Koru, Fort Ternan, and Songhor, in Kenya. The volcanoes associated with the Rift Valley spewed forth highly alkaline ashes and lavas throughout the Miocene. When these weathered into soils or were transported by rivers, they produced highly alkaline depositional environments favorable to the preservation of bones and

Table 3.2 Characteristics of fluvial deposits.

Deposit	Characteristics
Channel fill	Coarse gravel at bottom with increasingly finer sediments higher in the deposit. Channel bottom may be scoured and eroded and overlaid by conglomerates; overlying sands often show ripple marks; silts are uppermost.
Overbank deposits (floodplain)	Silts or clays deposited in discrete pockets lateral to channel.
Oxbow lake	Gravel and coarse deposits at bottom, overlaid by silt or clay. Organic debris common. More silt and clay than in channel fills.
Point bar	Predominantly sand and grit with small amounts of clay and silt.

teeth. At Napak, Bishop (1980) reconstructed the following sequence of events in the preservation process:
1) Alkaline tuffs give rise to strongly alkaline soils as the area is colonized by plants and animals. Bones resulting from attritional deaths are covered by later eruptions of windborne ashes. Permineralization, the infiltration of minerals into the bones, occurs.
2) Calcium carbonate, in groundwater from a local spring eye, strengthens the calcified fossils and cements fragments of bones and tooth together.
3) The calcified, reinforced fossils are found today in a catchment area that receives the fossils eroded out of the original sediments. In shallow valleys, such as Napak IV, or flats, such as Napak I, the fossils are discovered by paleontologists.

CONTINENTAL MARGINS

Considerable deposition of sediments occurs on the edges of continents where rivers flow into the ocean, forming deltas, lagoons, tidal flats, and offshore sand bars that may preserve terrestrial and marine vertebrate fossils.

Continental margins record the interplay between the forces of the rivers bringing terrestrial sediments to the shore and the forces of the ocean, which redistribute those sediments. A delta is formed when a river carries more sediment to the ocean than can be immediately redistributed. (Deltas also form where sediment-rich rivers empty into large lakes, or other rivers, so the following explanation applies to that environment as well.) Progressively finer sediments are deposited as the river currents flow into the ocean. Levees are the raised banks on either side of the channel, made up of earlier overbank deposits. Sedimentation at the mouth of the river eventually raises the water level so much that it overflows the levees, creating distributary channels (Figure 3.4) across the silty levees. In essence, a delta is an alluvial floodplain building itself seaward. Clays and silts settle out of the water at the forward edge, in the area known as the prodelta. As the delta builds outward, it creates the following sequence of beds: on the bottom are the fine silts and clays of the original prodelta, overlaid by the muds and sands of the delta slope, which in turn are overlaid by the coarse

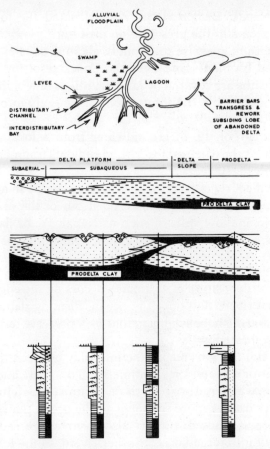

Figure 3.4 Aerial view of a bird's foot delta and vertical cross-section of typical deposits. (From Richard C. Selley, *Ancient Sedimentary Environments*, copyright © 1970, 1980 Richard C. Selley, reprinted by permission of the publishers, Cornell University Press and Chapman and Hall, Ltd.)

deposits of the distributary channels. Deltas migrate laterally as well, so such deposits are often cyclical.

Between deltas other sedimentary shoreline environments are created. Lagoons are shallow bays cut off from the ocean by offshore islands or sand bars that run roughly parallel to the shoreline. Clays and silts settle slowly out of these protected waters, forming a fine-grained substrate on and in which many marine organisms live. Their tracks and burrows can often be seen in lagoon sediments. The shoreline may build outward as a coastal alluvial plane because of the terrestrial sediments brought down by rivers and redistributed by ocean currents. Marshes, estuaries, and brackish swamps are common in such areas. The transition between this plane and the lagoon is marked by tidal flats, gently sloping deposits of clay, silt, and fine sands that are exposed at low tide and covered with water at high tide. Tidal flats, like lagoon bottoms, often show a great deal of bioturbation. In unusually salty lagoons or brackish swamps, evaporites may be precipitated. Offshore bars or barrier islands are accumulations of sand that may be derived from terrestrial sources, in which case terrestrial vertebrates may be found as fossils, as at Langebaanweg, South Africa.

INLAND BASINS AND TRAPS

Bishop (1980) recognized three sedimentary environments within this broad category:
1) Aqueous and semiaqueous environments such as deltas; lake margins or bottoms; channel lags; peat bogs, swamps, or tar pits; and springs or spring eyes
2) Caves, sinkholes, and other natural traps
3) Deserts, including the dune fields of hot deserts, and cold deserts or steppes

Aqueous and semiaqueous. Deltas that build outward into large lakes have the same characteristics as those at continental margins. Lake margins or lake bottoms, which can be grouped under the general heading of lacustrine environments, are often rich in preserved materials, including bones carried into the lake by rivers as well as bones of aquatic animals. If lake levels fluctuate over time, as did the ancient lake at Olduvai Gorge, Tanzania (Hay 1976), bones accumulate at the margins when the water is low and are covered over and preserved when the lake rises again. Such fluctuations may occur seasonally, but even broadly spaced fluctuations allow for preservation. Lake margin habitats seem to have been attractive to hominids, as at Olduvai, Tanzania, and at Olorgesailie and East Turkana, Kenya, as well as to other water-dependent species. Lacustrine deposits are characteristically fine-grained deposits of clays or silts that occur in many fine laminae. Near the margins and mouths of rivers, coarser materials may be deposited. Areas that are exposed to air by the lowering of the lake may show dessication cracks and bioturbation from aquatic species. Strand lines may be detectable from the accumulation of coarser particles aligned roughly parallel to the shoreline (see Hill and Walker, 1972, on Bukwa, Uganda). All fossiliferous lacustrine deposits are likely to include substantial proportions of both aquatic and terrestrial species.

Channel lags, which differ only slightly from channel deposits, are coarse deposits of gravel, bone fragments, teeth, and other very dense particles that remain in suspension only with high current velocities. They may be left as a lag deposit if the current cannot transport them, or they may be transported and then deposited farther downstream. Thus lag deposits are the equivalent of the bottom coarse layer of channel fills. Many of the fossiliferous deposits from the Omo River, Ethiopia, are channel lag deposits (Butzer, 1971; Bishop, 1976, 1980) with barren overlying deposits. Many of the Omo fossils are fragmentary, and the less dense or more fragile body parts are poorly represented, as is characteristic of fossils in channel lags.

Peat bogs, swamps, and tar pits may be sites of unusual preservation events, with remarkably complete remains. Peat bogs seem to effectively tan bodies in the same manner that hides are tanned to make leather. Complete skeletons with intact soft tissues are reported from peat bogs in Denmark (Glob, 1954) and other areas of Europe. These preserved bodies are only several thousand years old; it is unfortunate that the more intriguing, extinct species have not been so preserved. However, unusually complete skeletal remains without soft tissues have been found in swamps, such as the *Oreopithecus bambolii* partial skeleton from Tuscany, Italy. This individual, although one of the most complete fossil primates known, suffered extensive postmortem crushing. Although the decaying vegetable matter in swamps and peat bogs protects skeletal elements from prefossilization damage, it has little influence over postfossilization events. In tar pits, liquid asphalt acts in much the same way as water. The spectacular preservation of animals at La Brea, California, attests to the ability of tar to preserve and protect bones.

Springs occur where the water table intersects the ground surface, often on the slopes of valleys or where impermeable layers of rock deep underground cause the water table to rise when rainfall increases. Artesian springs occur where groundwater is trapped in a permeable layer of sand or gravel between two impermeable layers, often of shale. At a point where the permeable layer has access to the surface, the system is charged with rainwater, which is forced into the reservoir defined by the impermeable layers. As more and more water accumulates, the pressure increases, forcing the water up through any hole or flaw in the uppermost impermeable layer (Figure 3.5). Such openings are called spring eyes, and preservation of skeletal remains in or near them is apparently dependent on an alkaline environment. A typical sequence of beds for a spring eye includes a permeable sand or gravel layer bracketed by shales; the sediments enclosing the fossils may be clays (Figure 3.6). One well-studied spring eye at Boney Spring, Missouri (Saunders, 1977), is notable both for the large number of mammoth fossils and for the unusual, ring-shaped distribution of fossils around the spring eye.

Caves, sinkholes, and other natural traps. These may be excellent environments for the preservation of bones. Animals may literally fall into caves or sinkholes and die, or their remains may be trans-

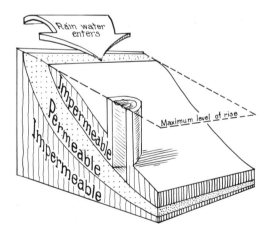

Figure 3.5 An artesian spring. Groundwater enters a permeable layer and is forced upward through a hole or defect. (Drawing by Dave Bichell.)

Geologic Setting / 57

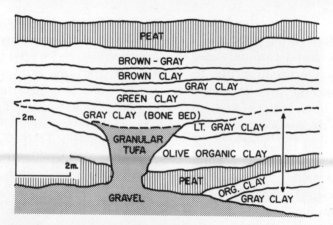

Figure 3.6 A simplified cross-section of the deposit at Boney Spring, Missouri. Notice the permeable gravel layer bracketed by shales; this is a typical spring eye deposit. (After Saunders, 1977; by permission of the Illinois State Museum.)

ported there by predators (human or nonhuman), water, or the general downhill creep of surface debris. Sediments may then accumulate and bury the remains. An especially well-known example of such a depositional environment is the group of South African limestone caves from which the australopithecine fossils are known.

Brain (1958, 1967) has meticulously reconstructed a typical sequence in the development of such cave deposits, (pictured in Figure 3.7) as follows:

a) An underground cavern develops as groundwater dissolves away limestone in the phreatic zone (also called the zone of saturation) below the water table.

b) The water table drops, bringing the cavern into the zone of aeration, or vadose zone. A second cavern begins to develop below the first. Rainwater seeps downward from the surface through natural fractures or joints in the limestone, enlarging these cracks by solution. The dissolved calcium carbonate is redeposited as travertine or flowstone on the roof and floor of the upper cavern.

c) Further enlargement of the joints provides the upper cavern with a direct opening to the surface. Soil, wood, bones, and

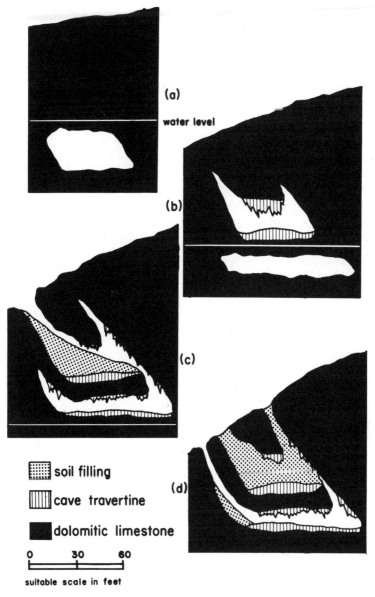

Figure 3.7 The sequence of development of a limestone cave deposit. (From Brain, 1967; reprinted from *Background to Evolution in Africa* by W. W. Bishop and J. Desmond Clark, eds., by permission of the author, the editors, and the University of Chicago Press. © 1967 by the Wenner-Gren Foundation.)

Geologic Setting / 59

other surface debris are carried into the cave by surface runoff or by slumping of the sediments; these form an irregularly shaped talus heap. The upper cave may also be inhabited by hominids, mammalian predators, owls, or other animals that contribute bones and debris to the heap. Blocks of limestone and travertine fall from the roof as rainwater continues to enlarge the upper cavern. Continued lowering of the water table brings the lower cavern into the vadose zone.

d) As the upper cavern fills with debris, the lower cavern develops openings to the surface, the upper cavern, or both, and also begins accumulating debris.

The strongly alkaline nature of the sediments is an important factor in preservation. Typically, the stratigraphy of such caves is complex, because items fall onto the talus heap and come to rest in unpredictable sequences and positions. The situation is further complicated by the possibility that the external opening may open and close many times, with periods of no deposition while it is closed. Bones preserved in such caves are often broken by postmortem events, and partial skeletons are rare; bones usually come into the cave in disarticulation.

Sinkholes are depressions, usually in limestone areas, with no outlet for water. They are often caused by the collapse of underground caverns. Some sinkholes are large, vertical-sided shafts of considerable depth, others are funnel- or saucer-shaped. Rainwater drains into sinkholes from the surface and percolates down through joints in the underlying rocks into underground caverns. Deep sinkholes, such as Crankshaft Cave, Missouri (Parmalee, Oesch, and Guilday, 1969), may serve as natural animal traps, with skeletons deposited in the debris at the bottom of the shaft (Figure 3.8). Runoff may also carry bones and debris into sinkholes, forming poorly sorted talus heaps like those in limestone caves. Depending upon the mode of accumulation, the preserved remains may range from complete or nearly complete skeletons to fragments of bone.

Deserts. With little rainfall and vegetation, water erosion and transport of sediment in deserts are marked when rainfall does occur. However, in such environments active sedimentation usually occurs through settling of windborne particles. Desert soils

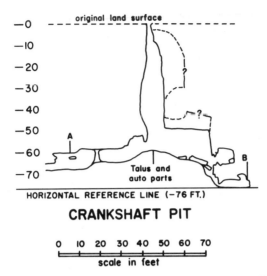

Figure 3.8 Cross-section of Crankshaft Pit, Missouri, a deep sinkhole. (From Parmalee et al., 1969; reprinted by permission of the Illinois State Museum.)

are characteristically poor in organic content and may be merely a hard surface of rock or pebbles (a pavement) with a wind- and water-winnowed lag deposit, called a deflation lag, on top (Butzer, 1976). These processes often concentrate previously buried stone tools and fossils. Wind erosion is especially important in desert areas. Evaporites may precipitate in shallow, saline lakes or ponds. If soluble carbonates, originally carried into the subsoil by rainwater, are drawn upward by capillary action during droughts, alkaline salts, including calcium carbonate, may be deposited just below the rock surface or in fissures (Butzer, 1976). These salts may cement soil particles together to form caliche deposits that slow the infiltration of water into the soil, thus increasing runoff and erosion in the future. Formation of a caliche deposit depends upon the presence of calcium and other soluble minerals in the source rocks from which the soil particles are derived. Bones or teeth covered by such alkaline soils are likely to be calcified and preserved.

Steppes, or deserts in areas where winters are extremely cold, may preserve the remains of animals in the permafrost (perpetually frozen layers). Examples are the frozen mammoths recovered from the Siberian permafrost. As with peat bogs, whole or nearly whole bodies are preserved with intact soft tissues. Unfortunately, none of the frozen animals found to date is more than several thousand years old.

Fossilization and Sedimentary Environments

In this brief review of depositional environments, I have referred to the characteristics of fossils found in various environments. Behrensmeyer (1973, 1975; Behrensmeyer and Dechant-Boaz, 1980) has focused on the differential preservation of skeletal remains in different environments associated with water. She found that the ratio of teeth—usually the densest element in the skeleton—to vertebrae—one of the least dense elements—in a deposit is broadly indicative of the environment of deposition. She observed that channel fills or lag deposits typically show a very high tooth to vertebra (T/V) ratio, as well as an absolutely high number of teeth and other dense elements. Many of the fossils found in such deposits have been rolled, broken, and abraded. Different depositional environments show different proportions of skeletal elements. In deltaic or lacustrine environments, vertebrae (and other light elements such as phalanges) are considerably more common, and the T/V ratio is lower. Fossils are often fresh and unabraded, suggesting minimal transport. These assemblages are apparently little altered by postmortem events.

Behrensmeyer also observed that skeletal representation in floodplain deposits resembles that in channels, but the matrix of surrounding sediment is usually fine-grained. She suggests that although hydrodynamic sorting has apparently not occurred, sorting by survival may be important. That is, fossils in floodplain deposits show signs of weathering and of having been chewed; presumably many fragile bones have been destroyed by nonfluvial processes prior to deposition.

Other environments also have characteristic patterns of fossilization. Whole or nearly whole skeletons are most likely to be preserved in swamps, bogs, tar pits, and steppes. Catastrophic death and rapid burial in other environments—falling into a natural trap, drowning—may also result in such preservation. In general, more complete skeletal elements are found in sedimentary environments in which postdepositional disturbances are less likely to occur and in which depositional events are less likely to cause destruction or damage. Thus lagoonal and lacustrine deposits can be expected to contain more complete remains than channel deposits.

It should be clear that there is a complex relationship among sedimentary environments, modes of accumulation and breakage, and the differential preservation of skeletal elements. Few hard and fast rules exist that can be applied in all situations. Geologic and sedimentary evidence must be integrated with other types of data to produce sound taphonomic reconstructions. Also, it is important to remember that present environments may bear little or no resemblance to the past environments in the same area. Nearly all fossil sites are found because they are presently in erosional areas, yet nearly all were formed because they were in depositional areas.

4 Spatial Distribution of Fossils in Sediments

The various agents that cause fossils to be concentrated in a particular area, such as predators, hominids, or water currents, produce characteristic patterns of spatial distribution of the fossils in sediments. Collecting and analyzing data on spatial distribution can reveal much about both the paleoenvironment and the events that led to the formation of the site. Concern with such data and its analysis is relatively recent in paleontology, and this information was not recorded, unfortunately, for most paleontological sites excavated prior to the 1960s; archaeologists, however, have more often collected spatial distribution data. This historical accident means that the analysis of the spatial distribution of fossils excavated in the past is often difficult, if not impossible.

Types of Spatial Distribution Data

The most obvious variables used to express spatial distribution are those that record basic position: a set of coordinates representing the position of the fossil in three dimensions relative to a fixed datum point. Alternatively, position may be more loosely recorded by noting which square of a predetermined grid includes the fossil. Collection of coordinate data is painstaking, slow, and boring, but the information gained by analyzing such data may be tremendous. The advantages of this method of recording spatial distribution are that it frees the excavators from the attempt to excavate at a steady rate in arbitrarily defined squares and makes it easy to follow fossil concentrations. However, grids may be important in designing a mathematically adequate sampling strategy that lessens the chances of missing such concentrations, especially when large areas are to be sampled, as at many archaeological sites. To some extent the mode of recording spatial distribution must reflect the type of site: an archaeological site, in which hominids have determined the distribution of artifacts, may require an initial period of random sampling to locate major features; at paleontological sites the most effective strategy may be to determine the geological strata or features associated with the bones and then excavate in the appropriate areas.

Once gathered, spatial distribution data may be analyzed in many ways to determine the density of fossils, their location relative to other features, and the patterning of their distribution. Analysis of the data may focus on four different factors, depending on the data available and the aims of the investigation: (1) horizontal distribution of fossils in sediments; (2) vertical distribution; (3) orientation of fossils; and (4) dip of fossils. Each of these can reveal different sorts of information about the paleoenvironment and the site formation events. Ideally, all four foci should be considered and the results integrated to make a comprehensive and detailed interpretation of the site.

HORIZONTAL DISTRIBUTION

Analysis of horizontal distribution is useful primarily in reconstructing the patterns of concentration that existed when the site was formed. Such analyses were used to locate the concentrations of bones in the channels at Fort Ternan and to define the roughly circular areas of high bone densities at certain sites at Olduvai. In these cases, the location and density of bones on one or more horizontal layers were examined, and the position of each bone on north–south and east–west axes was measured. From such data, both conventional site plans and density contours can be drawn.

On a broader scale, analysis of horizontal distribution can be coupled with analysis of faunal representation (see Chapter 6) and facies changes to delineate local habitats. A good example of such an analysis is that by Behrensmeyer (1975) at East Turkana. To determine the proportions of different types of animals and different skeletal elements that could be expected in different environments, Behrensmeyer identified all fossil fragments in at least twenty squares measuring 10 meters by 10 meters in each of three broadly defined environments: channel, floodplain, and delta. Sampling localities were selected to minimize possible contamination of the squares with material from other horizons. All bones larger than 5 centimeters in maximum length were identified to taxon and skeletal element, if possible, and their relative abundance calculated according to the number of squares in which a given element of a given species was found. The square frequency is derived by dividing the total number of squares in the environment where a particular bone or taxon was found by the number of squares sampled in that environment. This innovative method of analysis, especially useful in sampling large areas, avoids the problem caused by fragmentation of a single bone or tooth into several recognizable pieces, which inflates the counted number of individuals. Also this method helps ensure that a single disarticulated carcass will not distort the relative abundance data.

By examining data on faunal and skeletal representation and that on spatial distribution, Behrensmeyer was able to establish several general principles relating paleoenvironment to the spatial distribution of fossils. She found in general that the relative abundance of vertebrate classes in the fossil assemblage reflected their original abundance and proximity to sedimentary environments. There was a greater diversity of terrestrial mammals in terrestrial environments, although both terrestrial and aquatic species were found in both aquatic and terrestrial environments. The remains of aquatic species were more abundant in aquatic environments, but the abundance of terrestrial species was comparable in aquatic and terrestrial environments. Finally, animals that were more abundant in life were represented by a wider range of fossil skeletal elements with different hydrodynamic properties than those animals that were originally less abundant. Although these conclusions cannot be expected to hold true for all environments and all species, they are probably broadly applicable.

VERTICAL DISTRIBUTION

The analysis of vertical distribution of fossils in sediments can be on either a large or a small scale. On a large scale, differences in both the absolute and relative numbers of different species at different levels may reflect changes in the environment over time. For example, a mixed forest-savannah fauna might, over time, show increasing numbers and diversity of forest species. This might indicate that the local environment was becoming more heavily forested, with open country (and open-country animals) becoming scarce. To detect such large-scale changes through analysis of vertical distribution of fossils, it is necessary to find large exposures of several fossiliferous beds, representing a considerable time period. Otherwise, random fluctuations in the local faunas sampled may produce apparent differences in faunal composition over time.

It is also important to consider the units of comparison: are the data the raw numbers of specimens per species in each stratum, or are they the minimum number of individuals (see Chapter 6) represented by those specimens? Or does the comparison focus on changes in the number and diversity of species within higher taxonomic groups? Depending on the interests of the investigator, any of these comparisons might be valid. However, it is crucial that the same unit, derived in identical ways, be used for each assemblage being compared.

Small-scale analyses of vertical distribution may use data from a single horizon or perhaps a few horizons, representing local and relatively rapid depositional events. An example is the analysis of the data from Fort Ternan, showing the concentration of bones in a network of channels 1 to 2 meters deep over an area roughly 50 by 35 meters (Shipman, 1977). Such intensive analysis may reveal many details about the events during site formation that were responsible for the observed distribution of fossils.

If the aim of the analysis is to investigate broad ecological changes over a time span, it may be sufficient to know only the stratum in which each fossil was found. An advantage to this level of analysis is that by examining the matrix adhering to the fossils, it is possible to trace even those fossils that have eroded out of the sediments to their original horizon, and the relevant data about vertical distribution can be reconstructed. Therefore, the analysis need not be restricted to excavated specimens. If the purpose of the analysis is to reconstruct site formation events or to trace the agents of concentration that acted within a particular horizon, the raw data on vertical distribution must be correspondingly precise.

Prior to the analysis, the coordinates representing vertical position (depth) must be corrected for distortion produced by the general dip (degree of declination) of the beds. If the beds have dipped, bones originally laid down on the same level may be found at substantially different depths (Figure 4.1). For example, over a distance of 5 meters, a modest dip of 4 degrees causes a maximum depth distortion of 20 centimeters.

ORIENTATION

The preferred orientation, or dominant compass direction, of fossils in an assemblage may indicate life positions, bioturbation (dis-

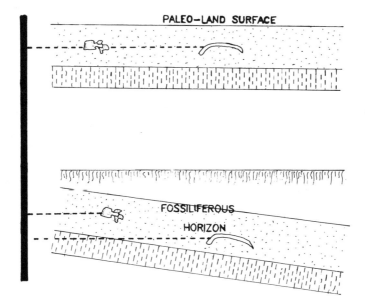

Figure 4.1 Postdepositional dipping may distort depth relationships. *Top*, two bones are buried at the same level within a horizontal stratum. *Bottom*, postdepositional dipping of the stratum causes one bone to be much deeper than the other relative to the modern land surface, although their depth relative to the top of the stratum is unchanged.

turbance of the sediments by living species), the axis of current flow, or the axis of wind flow. The orientation of the long axis of fossils in a horizontal plane has been recorded at several sites, including the Valentine Formation in Nebraska (Voorhies, 1969); Bukwa, Uganda (Hill, 1975; Hill and Walker, 1972); Rusinga (Andrews, Van Couvering, and Van Couvering, 1972); Olduvai (Hill, 1975); Omo, Ethiopia (Johanson et al., 1976); Boney Spring, Missouri (Saunders, 1977); and others.

With all but the smallest vertebrates, a preferred orientation is most probably caused by water action, in the absence of evidence to the contrary. Orientation according to life position is most common among sessile organisms and is therefore of more use to invertebrate paleontologists than to vertebrate paleontologists (Ager, 1963). If bioturbation occurs in a consistent direction for some reason, it may produce a preferred orientation. In that case, close study of geologic sections should reveal other evidence of bioturbation, such as reworking of sediments or visible sections of tunnels, burrows, or tracks. If current or wind has caused a preferred orientation of the fossils, this may be deduced from the existence of waterlaid or aeolian sediments, respectively. In some instances, the course of a small waterway can be reconstructed from data on the concentration and orientation of fossils. Generally, geologic study clarifies whether water or wind was responsible for a preferred orientation. In ambiguous situations, it may be useful to consider the size of the oriented specimens, since normal wind speeds are insufficient to move any but very small bones.

Unfortunately, in assessing the significance of orientation data, there are serious theoretical difficulties that have not yet been resolved, as indicated by the following questions:

1) What is an adequate sample of bones for determining preferred orientation?
2) Does the sample differ significantly from a random distribution of orientations?
3) How is deviation from the preferred axis to be measured?
4) How much deviation from the preferred axis can reasonably be expected to occur without implying a change in direction of the current?
5) What current velocity is necessary to produce a preferred orientation?

6) What patterns of orientation are produced by different environments such as channels, deltas, and lake margins?

Most of these questions cannot be answered fully without further research, observation of modern situations, and continuing experimental studies using artificial flumes. Until further experimentation provides an answer to the first question, it is suggested that a sample containing fewer than seventy-two bones be regarded as inadequate. This size sample permits bones to be placed in 10-degree units by grouping all bones with orientations between 1 degree and 11 degrees, then all those with orientations between 12 degrees and 21 degrees, and so on. This grouping method allows for an error in measurement of up to ±5 degrees. If fewer than seventy-two bones are used, there will be an average of less than two to three bones in each of thirty-six 10-degree units, and the results will be dramatically influenced by a single measurement error. Once grouped, orientation data are presented in a rose diagram, constructed of thirty-six radially arranged wedges. The length of each wedge shows the number of bones within one 10-degree unit.

Sometimes mirror-image rose diagrams are constructed, in which bones oriented along the same axis are grouped into eighteen 10-degree units. Thus, all bones lying between 1 degree and 11 degrees would be grouped with all bones lying between 181 degrees and 192 degrees. In such rose diagrams, there is an axis of symmetry, and the two halves of the diagram are mirror images of each other. Mirror-image rose diagrams are useful because they exaggerate the axis of preferred orientation, although they conceal the preferred direction if there is one.

Mardia (1972) proposed a partial answer to questions 2 and 3, by treating orientation data as if they were metrical. He suggested that rose diagrams be divided into one or two segments and "unpeeled" (see Figure 4.2). This method allows the researcher to use standard statistical tests for deviation from a normal curve.

There are several caveats to be observed here, however. First, the placement of the divisions may alter the resultant image considerably. A unimodal distribution, divided within its peak, will appear to be clearly bimodal. Second, a specimen oriented along an axis can be measured by two opposite directional readings; for example, a bone oriented at 1 degree is also oriented at 181 degrees. If the items being oriented are equally stable, with either

Figure 4.2 "Unpeeling" a rose diagram to transform it into a histogram. (Drawing by Dave Bichell.)

end downflow, the result is a preferred *axis* of orientation rather than a preferred orientation. In this case, the two segments suggested by Mardia would show patterns insignificantly different from one another only if the divisions were made at a 90-degree angle to the preferred axis. In such a case, a mirror-image rose diagram, in which the number of specimens at a given orientation is added to the number at that orientation plus 180 degrees, is a realistic representation. But when the mechanical stability of the oriented items varies according to which end is downflow, mirror-image rose diagrams unduly emphasize the axis of flow while concealing the direction.

It is relevant to this analysis that Voorhies (1969) reported that long bones orient with their heavier ends upstream. In this case the entire 360-degree spectrum would represent a single normal curve wrapped around a cylinder as Mardia suggests, although mistakes in measuring azimuths may produce a second, artificial peak at 180 degrees from the true peak of the curve. However, there is an important difference between normal curves and orientation data. Normal curves are found in the distribution of variables that have no theoretical upper limits to their values, such as length of femora. Directional data are both theoretically and practically limited at 360 degrees. A bone being oriented by a current cannot rotate through 360 degrees without returning to its original position; rotating through 360 degrees is, therefore, the same as not rotating at all, in terms of measurable difference in orientation. Directional data are basically cyclical, not metrical. Therefore, directional data violate the basic premise of the statistics Mardia proposes using: that the data might approximate a normal curve. If the data cannot approximate a normal curve, it is obviously inappropriate to use statistical measures of deviation from a normal curve.

These are theoretical problems, whereas questions 4, 5, and 6 are mechanical ones. A consistent flow will orient many bones, shells, or other items along the general axis of flow, and this is the reason for analyzing data whose theoretical significance is unclear. Voorhies (1969) found that long bones tend to orient themselves with their long axes parallel to the current flow unless the bones are partly out of the water, in which case there are transverse alignments as well. Bones are most stable in whatever orientation minimizes the surface area subject to drag from the current. Once they are in this orientation, a higher critical shear stress is necessary to initiate further movement.

How far the orientation of the bones deviates from parallel to the axis of current flow depends on the density and shape of the bones; their settling velocity and critical shear stress; the velocity of the flow; the morphology of the surface on which the bones are being oriented; and the effects of other particles or bones in the current. The role of these various factors in determining orientation has not yet been adequately determined for modern bones, so it is unwise to speculate on their role in ancient environments. Much work is also needed to determine the orientation patterns produced in different modern environments with different current velocities; our knowledge of these patterns is, at best, rudimentary. In short, many of the theoretical difficulties in assessing orientation data remain unresolved. However, use of the procedures described below, developed for the Fort Ternan data (Shipman, 1977; Shipman, Walker, Van Couvering, Hooker, and Miller, 1981), will do much to lessen the confusion.

First, orientation data are grouped into eighteen 10-degree units, as described earlier for constructing mirror-image rose diagrams. Grouping by larger units would obscure orientation peaks that fall off sharply, and it is precisely those peaks that are most useful in analyzing orientation data. Once the data are grouped, the basic pattern and the strength of the peaks of preferred orientation must be assessed. Three basic models of orientation patterns have been constructed (Figure 4.3), approximating three different environmental situations. Model I shows strongly oriented particles with a single, major preferred axis of orientation. This type of distribution results from strong current action on uniform particles. The peaks step off sharply and, because this is a hypothetical

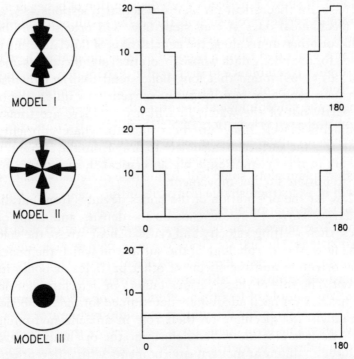

Figure 4.3 Orientation patterns: model I—a preferred axis of orientation becomes a bimodal histogram; model II—two axes of orientation at right angles to each other become a trimodal histogram; model III—randomly oriented particles in a circular distribution become an even histogram.

model, symmetrically. It may be argued that the strongest possible orientation of this type is one in which 50 percent of the bones fall in each of two wedges at 180 degrees from one another; however, this situation is most unlikely to occur, and any rose diagram of real orientation data will differ from it substantially. Model I is probably a more realistic simulation.

Model II is a rose diagram of strongly oriented particles with a major axis of orientation and a minor axis at right angles to it. This is the basic cross pattern reported by Voorhies (1969) in experimental flumes in which the bones were partially out of the water. Like model I, this is not the most extreme theoretical case, which would be only four wedges, each containing 25 percent of the bones.

Model III shows a perfectly even distribution of bones in a rose diagram. A perfect circle such as this is the result of a completely random orientation of bones or particles. Figure 4.4 shows rose diagrams generated using a random number program. These bear a reasonable resemblance to the circular distribution, even though the circle represents the extreme. They also resemble rose diagrams in the paleontological literature that have been considered to show preferred orientations, which highlights the importance of analyzing any apparent preferred orientation, regardless of the investigator's intuitive assessment of the data. Of course, given enough random plots, some will mimic true preferred orientations and will be indistinguishable from orientations caused by some aspect of site formation or deposition.

The three models can be used to study orientation data from real situations. The first step is to convert each rose diagram into a histogram of percentage frequency of orientations. This conversion standardizes samples of different sizes although, as I have mentioned, any sample of less than seventy-two observations may be inadequate. Making the orientation data into histograms does not involve the same error made by Mardia (1972), that of assuming directional data to be metrical. These histograms may be regarded as plots of cyclic or periodic phenomena, thus preserving the true nature of the data. Model I becomes a bimodal distribution. The histogram begins in the middle of the preferred orientation peak, because that is the axis of symmetry of a mirror-image rose dia-

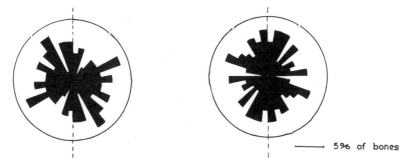

Figure 4.4 Rose diagrams generated with a random number program sometimes appear to show preferred orientations, but these do not differ significantly from a circular (random) distribution.

gram from real observations. Model II, the cross, becomes a trimodal distribution. Model III, the circle, becomes a histogram with 5.5 percent of the particles in each unit.

To compare real orientation data with these models, convert the observed data into a histogram showing percentage frequency of different orientations and superimpose this histogram on each model in turn. Whenever the histogram for the real data falls below that of the model, the number of units of difference is noted. The sum of these differences gives a percentage difference between the model and the data. In each case, a 0 percent difference constitutes identity; for models I and II, the circle represents total lack of identity. Percentage differences greater than 33.3 percent are judged so substantial that the data cannot be considered to "match" the model. Data that differ by more than 33.3 percent from both models I and II and by less than 33.3 percent from model III are considered random.

There are several advantages to using these models. First, this method involves no questionable or statistically invalid assumptions. Second, it provides a comparative measure of resemblance between sets of orientation data. Third, the causes of the different distributions in the models are at least partially known. The even distribution, as in model III, may result from the lack of any consistent influence on orientation. However, such a distribution has been documented at Boney Spring, Missouri, by Saunders (1977), as shown in Figure 4.5. The large number of bones at this site are in a nearly complete ring around a spring eye, which was apparently responsible for their preservation. The continuous flow of water up through the spring has oriented the bones nearly evenly, so that the distribution differs from random by only 8 percent. In this instance, then, an apparently random orientation of bones is actually a special case of radial orientation, which would not have been apparent if the geological evidence had not been analyzed in conjunction with the horizontal distribution and orientation data.

DIP OF FOSSILS

Orientation data record the position of a bone in a single plane only (Toots, 1965a:220). The dip (degree of declination) of bones provides another important description of their position. Unfortu-

Figure 4.5 Orientation of 517 fossils from Boney Spring, Missouri. (Data from Saunders, 1977.)

nately, the factors contributing to dip are not well understood. There is neither good experimental nor good comparative data from many fossil assemblages. Such data as there are indicate that low dips (0 to 10 degrees) predominate, as these nearly horizontal positions are most stable. High dips (70 to 90 degrees) in large numbers of bones, especially if the bones are of moderate to large size, suggests a special depositional environment. Voorhies (1969) interprets the presence of many high dips in large bones as evidence of extremely rapid deposition under flood conditions. Voorhies attempted to produce comparable dips in an artifical flume but could not, even at velocities up to 150 centimeters per second. However, high dips do not always indicate deposition under flood conditions; Hill and Walker (1972) have attributed the high dips at Bukwa to trampling rather than to a high-energy depositional environment.

In some instances, the dip of the beds may distort the dip and orientation of fossils as well as their position. Therefore, it is necessary to correct these figures prior to analysis. For example, if a bone is deposited in a horizontal position and, subsequent to lithification, the beds themselves dip 10 degrees, the bone will have an apparent dip of 10 degrees as measured in the field. As with the dip of bones, the dip of the beds has both magnitude and direction, which can be calculated from measurements of the level of the beds across a wide area. The correction needed on the apparent dip of the bone depends on its orientation relative to that of the direction of dip of the beds (Figure 4.6). For example, a bone that dips with the beds (one oriented between 300 and 360 degrees) needs the maximum correction (−4 degrees). However, as beds dip, there is an axis at right angles to the dipping plane that remains horizontal, called the strike. This can be most easily visualized if the beds are imagined to be a stack of papers, tilted so that their long axis slopes downward; the short axis is then the strike. Therefore, a bone that dips along the strike of the beds, which by definition has remained horizontal, needs no correction. Occasionally the correction process may produce a negative dip. In those cases the orientation is altered by 180 degrees to produce a positive dip in the opposite direction.

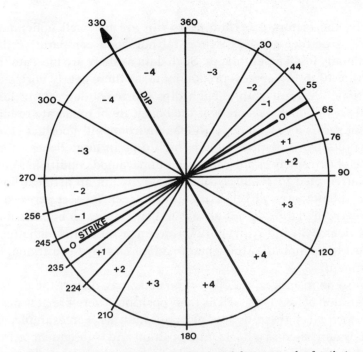

Figure 4.6 Dip correction diagram developed for use at the fossil site at Fort Ternan. The recorded dip of each fossil is corrected by the factor within the wedge corresponding to the orientation of that fossil.

Spatial Distribution and Paleoenvironments

Analysis of the spatial distribution data may take the form of a search for patterning, from which deductions about paleoenvironments and site formation events can be made, based on the type and clarity of pattern found. The different factors that are known to produce patterning in the spatial distribution of fossils are: (1) geologic events or conditions, such as hydraulic forces, lava or mud flows, and erosion; (2) biological events, such as activities of hominids; life habits, such as burrowing or frequenting river banks or lake margins; mode of death; and actions of other species after the death of the animals; and (3) temporal events, such as changes in species abundance, shifts in ecological zones, evolutionary changes in species' adaptations, and climatic changes.

GEOLOGIC EVENTS

The analysis of spatial distribution patterns in reconstructing local geologic events can be illustrated with the data from Fort Ternan. An area of this site, known as the main quarry, was excavated by M. D. and L. S. B. Leakey and their workmen between 1959 and 1967, and an area known as FT-B was excavated by a team of taphonomists and paleontologists in 1974. During the Leakeys' excavation, detailed site plans were drawn whenever a substantial number of fossils was exposed. One such site plan is shown in Figure 4.7. A grid of one-foot by one-foot squares was measured onto the exposed surface and then the position of each bone in its square was drawn on the site plan by eye. Unfortunately, no vertical controls were established other than by major geological strata, nor was the exact position or orientation of any fossil recorded. Nevertheless, these site plans represent a high standard of documentation of spatial distribution data for the time. In contrast, the 1974 excavators, who planned the project specifically to gather taphonomic data, measured and recorded the *in situ* positions of all fossils (N = 1,124) relative to a fixed datum point.

Because of their higher level of accuracy, I will discuss in detail only the data from the 1974 excavation, to show how such an analysis can be done (data from Shipman, 1977). There were three aspects to the analysis:

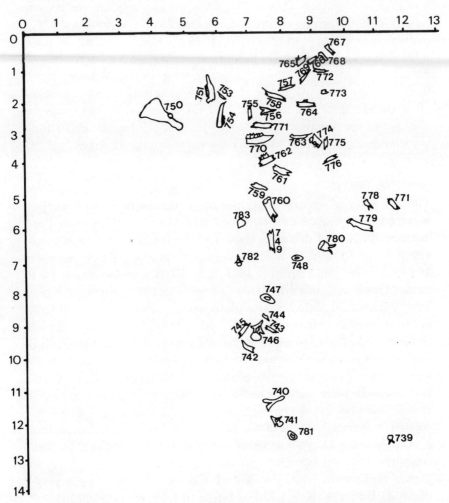

Figure 4.7 Fort Ternan: typical site plan from the early excavation in the Nzuve Trench. The number next to the drawing of each bone is the catalogue number.

1) By horizontal layers progressing down through the excavated area
2) By vertical cross-sections through the excavated area
3) By fossil density, using contour lines

A three-dimensional grid was superimposed on the coordinate data, dividing the excavated area into blocks for purposes of analysis. This procedure may seem to mask the additional accuracy of recording spatial distribution data in coordinates rather than squares. However, once coordinate data have been collected, they can be corrected for distortion and then analyzed using blocks or squares of any appropriate size. This block system can be adapted to many types of spatial distribution data and provides convenient units for computer analysis. Comparison of the density of concentration of fossils in different blocks is a straightforward measure of bone distribution. Further, particular horizontal layers or vertical sections can be singled out for intensive examination. If spatial distribution data are recorded initially only by documenting the square from which each fossil came, there is no possibility of correcting dip, orientation, or depth data, nor of varying the size of the units of analysis.

Of the 768 blocks in the Fort Ternan analysis, only 133 contained bones. This figure is artificially low because the outer boundaries of the grid were drawn around the outermost bones. Although this meant that unexcavated areas were included in the grid, it is a useful precaution except in cases where the researchers are very sure of the precise dimensions of the excavated area. In this case the excavated area was not even approximately cubic, for the excavation was deepest where the concentration of bones was highest (Figure 4.8). In looking at the distribution plots, one must bear in mind that the absence of bones on the peripheral areas is artificial; only the blank areas within the parameters of bone-bearing areas are significant.

In the 133 blocks containing bones, the number of bones per block ranged from 1 to 85. The mean number of bones per block was 8.5. Figure 4.9 shows that the high mean is produced by the 14 blocks that contain more than 20 bones each. Figure 4.10 shows that the bones in FT-B are concentrated in two major areas. In the older levels the bones lie between the northernmost extent of the excavation and an imaginary line roughly 120 centimeters south of

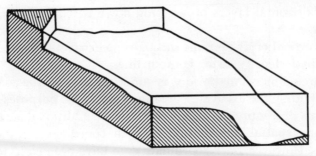

Figure 4.8 For analysis of coordinate data, using a block with dimensions that exceed those of the excavated area precludes accidental exclusion of any specimens. Unexcavated areas are shaded.

there. Bones are especially densely packed near the eastern edge of the excavation; all of the blocks with 20 bones or more lie here. This concentration corresponds to the location of a small gully or channel that could be clearly seen in cross-section prior to excava-

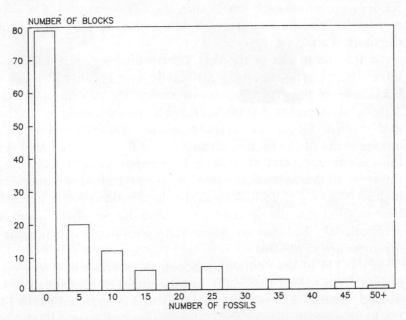

Figure 4.9 Histogram of the number of fossils per block in the FT-B area of Fort Ternan. Each block is 0.5 meters by 0.5 meters by 0.1 meter. 1,124 fossils were excavated from 133 blocks. The mean number of bones per block was 8.5.

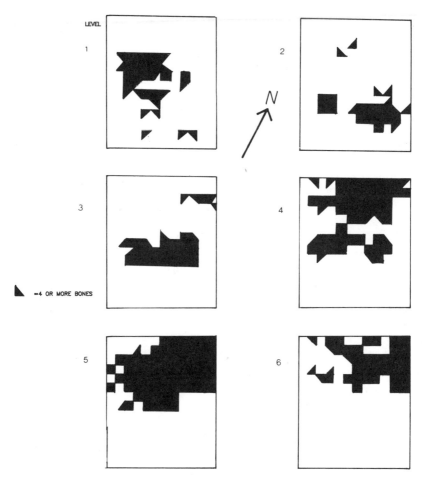

Figure 4.10 Distribution of fossils in FT-B on each of six horizontal 10-cm layers. Level 1 is uppermost and therefore youngest; Level 6, at the bottom of the sequence, is oldest.

tion. A secondary lateral concentration of bones appears in Level 3 and persists until the youngest levels, disappearing where the excavation begins sampling the overlying grits rather than the upper paleosol (ancient soil; Levels 1 and 2). Figure 4.11 shows another view of the concentration of bones, a vertical cross-section through the east face, where the excavation was deepest. Again, as in the horizontal data, both the gully or channel and the lateral deposit are evident.

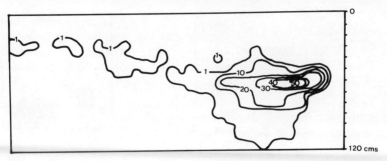

Figure 4.11 Vertical cross-section through the east face of the excavation of FT-B. The fossils are concentrated in a channel or gully.

In summary, both the horizontal and the vertical sections show a concentration of bones within a clearly defined channel. The bones are most numerous at the northernmost 120 centimeters of the excavation and between a depth of 40–60 centimeters. The concentration is densest at the eastern edge of the excavation, suggesting that the excavated areas of FT-B and the main quarry to the east of FT-B include most of the bones present in the upper paleosol. A similar analysis of the site plans of the main quarry reveals areas of bone concentration. Figure 4.12 shows contour lines superimposed on the excavated area of the main quarry, which

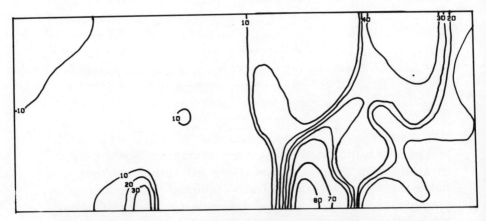

Figure 4.12 Fort Ternan, Kenya: contour map of fossil densities per square foot in the main quarry. Contour intervals are 10 bones per square meter.

shows more clearly than the other methods that the channel in FT-B is part of a network of small channels in which all the bones are concentrated.

Thus all three modes of analysis reveal a dense concentration of bones coinciding with a network of small gullies or channels. This dendritic pattern suggests that hydraulic forces may have been responsible for the distribution of the bones, but analysis of the orientation and dip data do not support this interpretation. The orientation data on 1,140 bones (Figure 4.13) show no significant preferred orientation when analyzed by 10-centimeter levels or when all orientation data are combined. The lack of preferred orientation persists even when the bones are grouped by skeletal element, by location relative to the channel or gully, by bone volume, or by degree of dip. Such a lack of preferred orientation rules out either water or wind as a major orienting influence and suggests that the assemblage as a whole is probably untransported and may represent a natural land-surface accumulation of bones.

The bones from the 1974 Fort Ternan assemblage show neither a consistent direction of dip nor imbricate structure (like the overlapping shingles on a roof) among the bones with high dips. Because imbricated particles tend to dip upstream, an asymmetrical rose diagram can be used to determine upstream direction when imbrication is present. But these bones are nearly evenly divided between a generally northeasterly and a southwesterly direction. In addition, the bones show a predominance of low dips (Figure 4.14): 76 percent have a dip of less than 20 degrees, indicating that most of the bones were in a mechanically stable position. Thus the dip data, like the orientation data, refute the hypothesis that hydraulic forces played a major role in the spatial distribution of these bones. Comparison of the orientation and dip of bones found in the channel with those found lateral to it reveals no substantial differences.

How, then, did the bones become concentrated in the channels at Fort Ternan if not by water currents? Geological evidence indicates that the area of the site was a small, ancient basin that underwent several cycles of increased wetness, when the basin was probably boggy, and dryness, when the sediments would become sufficiently dry for an occasional rain to cut small gullies such as

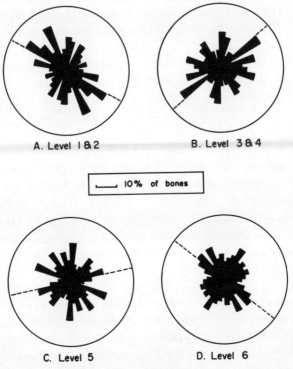

Figure 4.13 Mirror-image rose diagrams showing the orientations of fossils at different horizontal levels at Fort Ternan. Data from adjacent levels have been combined when the number of fossils in each is small. No level shows an orientation pattern that differs significantly from random.

those in which the bones are concentrated. The presence in the channel of ancient soils containing a jumble of unsorted and nearly randomly oriented bones suggests that the bones were slumped into the channels in a mud sludge that did not orient, sort, or transport them any significant distance. This interpretation is consistent with both the spatial distribution data and the geologic data. In this instance, then, the cause of the distinctive pattern of spatial distribution seems to be geologic.

HOMINID ACTIVITIES

Another influence on the spatial distribution of fossils is that of humans and their direct or indirect ancestors in hunting, butch-

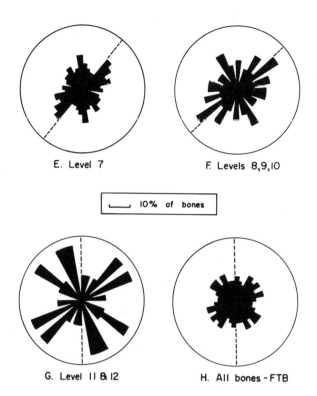

E. Level 7

F. Levels 8, 9, 10

10% of bones

G. Level 11 & 12

H. All bones - FTB

ering, and scavenging vertebrates, consuming their bodies and discarding their bones. Although the earliest signs of scavenging by hominids are difficult to distinguish, some patterns of breakage that are distinctly different from those caused by nonhominid forces have been detected at Plio-Pleistocene sites such as Olorgesailie, Kenya (Shipman, Bosler, and Davis, 1981).

But how does the spatial distribution of fossils reflect hominid activities such as butchering or consumption of animals? In order for hominid activities to be recognizable, they must produce patterns different from those of other possible influences on spatial distribution. Because tracing the activities of early hominids is so conducive to controversy, it seems preferable to use this explanation as a last resort when all other possible explanations have been ruled improbable.

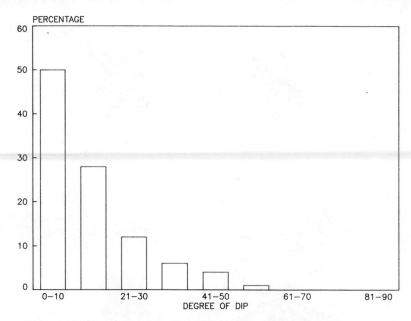

Figure 4.14 Histogram of the dips of fossils (N = 961) from the FT-B area of Fort Ternan, showing the predominance of low dips.

A single example will demonstrate the difficulties. Frison reported evidence of "deliberate stacking or piling of mammoth bones after butchering and processing by humans" (1976:728) at a purported Paleo-Indian site, dated to 11,200 ± 200 years, in northern Wyoming. Small concentrations, each containing many mammoth bones, were excavated in an old alluvium-filled arroyo (Figure 4.15). Two projectile points, one of which was unused, and a possible chopping tool were found in close association with the mammoth bones. Frison suggested that these concentrations represent deliberate stacking of mammoth bones by Paleo-Indians, but because of the paucity of the tools he did not feel that the stacking was directly related to butchering. Although his interpretation may well be correct, the figure reveals that the bones are concentrated at bends in the arroyo where whole or partial mammoth carcasses would probably be deposited if they had been moved by water currents. Only analysis of the representation of skeletal elements in each pile and of the sediments themselves could resolve

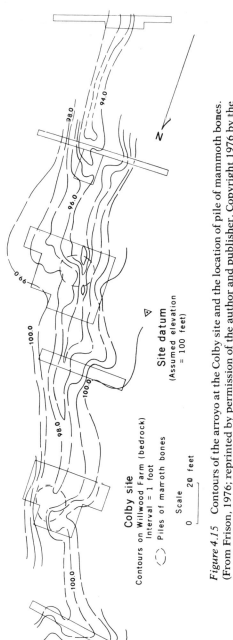

Figure 4.15 Contours of the arroyo at the Colby site and the location of pile of mammoth bones. (From Frison, 1976; reprinted by permission of the author and publisher. Copyright 1976 by the American Association for the Advancement of Science.)

the conflicting interpretations of these data. It must be remembered, however, that Frison is dealing with relatively recent people. His lack of caution in this instance may be because there is much documentation of similar behavior during historic times.

However, hominid patterns of bone concentration are recognizable in some circumstances despite the extreme antiquity of the sites. M. D. Leakey (1971) has documented several roughly circular areas of high bone concentrations at different Olduvai sites. One of these, at DK I level 3, is associated with a roughly ringlike distribution of stones, described as a stone circle (see Figure 4.16). Leakey suggests that these stones may have anchored branches used to

Figure 4.16 Olduvai Gorge, Tanzania: part of the site plan, showing the stone circle level at DK I, Level 3. (From Leakey, 1971; reprinted by permission of Cambridge University Press.)

form a crude hut or shelter, in a style like that seen today among various tribes. But similarly circular bone concentrations are also seen at FLK Zinj level, FLK North levels 1 and 2, and FC West, which do not have stone circles (Figures 4.17 and 4.18).

There is no geologic evidence of hydraulic activity at these sites, so it is highly unlikely that concentrations were produced by water currents. This conclusion is supported by the fact that the bones show no preferred orientation. Hyenas or other carnivores can also be ruled out as major agents of collection in this case, for four reasons. First, the fossil densities are too high and the skeletal representation too broad to be the remains left on a kill site by predators. Second, the faunal composition is so low in carnivores—which are often preferentially collected by other carnivores—that the concentrations are unlikely to represent a carnivore lair. Third, if the assemblage were a carnivore lair assemblage, the percentage of carnivore-damaged bones ought to be high, and it is not (Potts, personal communication). Fourth, there is no evidence in the immediate area of a burrow or cave that might have served as a lair. The bones are also too large to have been concentrated by wind action. In short, the only remaining agent of concentration is hominids, which have left abundant evidence of their presence in the form of stone tools.

This is not to say that hominids were responsible for the placement of every single bone at these sites; indeed, it seems likely that other forces probably shifted some of the bones from their original positions and may have brought some of the bones to the site. But hominids seem the most likely major cause of concentration in this case. Although it might be argued that the stone tools in themselves are sufficient evidence that the bones were collected by hominids, this is not true. Stone tools are hardly ever destroyed once they have been manufactured and may remain at or near the surface for many thousands of years, during which time they may become accidentally associated with bones of animals that died much later in time.

Can anything be learned from comparing ancient bone concentrations with those of modern humans? The use of modern analogies in interpreting ancient fossil sites is controversial, and the greater the time span, the more caution must be exercised, especially about matters of behavior that are not essential to life. Given

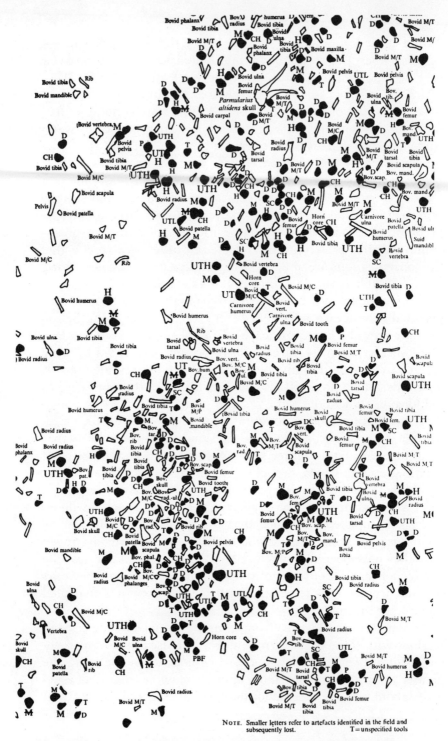

Figure 4.17 Olduvai Gorge: part of the site plan, showing roughly circular areas of high fossil densities at FLK North, Levels 1 and 2. (From Leakey, 1971; reprinted by permission of Cambridge University Press.)

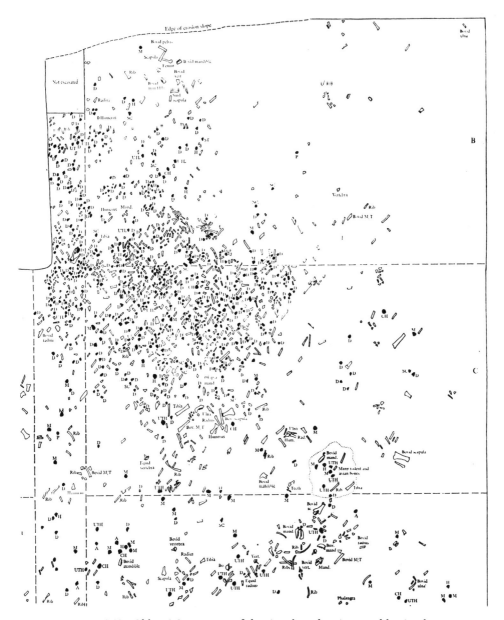

Figure 4.18 Olduvai Gorge: part of the site plan, showing roughly circular areas of high bone density at FLK, Zinj level. This site is interpreted as being a living floor. (From Leakey, 1971; reprinted by permission of Cambridge University Press.)

that roughly two million years have passed since the formation of the Olduvai sites, by hominids in a genus and species different from our own, extreme care must be taken in extrapolating from modern behaviors. Leakey (1971) is properly cautious in pointing out the similarities between the stone circle and the stone foundations of brush huts found today in some areas of the world. It is equally plausible to suggest that the stones at Olduvai DK I were used as a windbreak or for some reason that is absurd by modern standards (Figure 4.19). The stone circle, whatever its function, was not causally related to the concentration of bones, because very similar bone concentrations have been found at sites without

Figure 4.19 Possible uses of stone circles: to keep out animals; to anchor branches for a shelter; to contain a fire; as a primitive playpen; to enclose a garden; as a lizard-fighting arena. (Drawing by Dave Bichell.)

stone circles. It is possible, however, although it cannot be proved, that at these latter sites some other, less durable material such as thorn bushes served the same function as the stone circle.

It is apparent from these examples that spatial distribution alone will not always identify the agent of concentration. Sedimentological evidence, bone breakage analyses, and investigations of faunal and skeletal representation must be integrated with the data on spatial distribution to create a synthetic and sound interpretation.

ACTIONS OF OTHER ANIMALS

Many different sorts of animals may produce recognizable changes in the spatial distribution of bones. Porcupines, wolves, hyenas, leopards, and vultures, among other species, may collect bones of all sizes of animals in the process of feeding or in taking them to their lairs to gnaw. Smaller predators, such as foxes, jackals, eagles, hawks, or owls, may collect bones in regurgitated pellets or in scat. Harvester ants also collect bones, apparently to remove bits of meat from them. All of these behaviors are probably ancient adaptations and can be expected to have occurred in related or similarly adapted species in the past. In some cases the faunal or skeletal composition or surface characteristics of the bones may reveal the collector. For example, Figure 4.20 shows some of the characteristic gnawing patterns produced by porcupines. This is discussed further in Chapter 6.

Another important way in which living animals may alter the distribution of bones is through trampling. Gifford (1977) has observed that bones lying near water sources are often trampled by ungulates as they go to and from the water. When the bones are weathered, large bones—especially metapodia—are likely to be broken in distinctive ways by trampling. Smaller and denser bones are more likely to be trampled into the substrate without being damaged. Trampling in muddy lake sediments has been proposed by Hill and Walker (1972) as the mechanism by which a number of long, slender bones became nearly vertically oriented at Bukwa, Uganda. Geologic data from this site are inconsistent with the hypothesis that high-velocity currents deposited these bones in such mechanically unstable positions. Hill and Walker reasoned

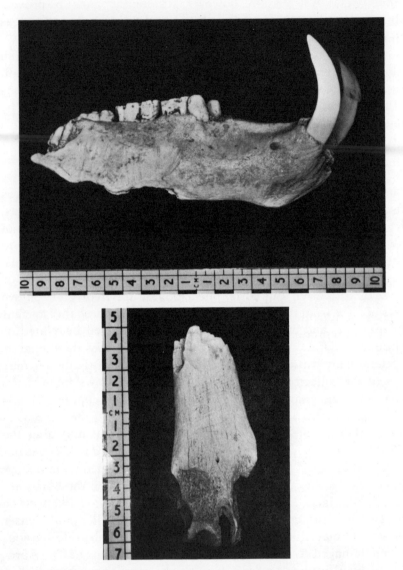

Figure 4.20 Porcupine gnawing marks on bones. *Top*, a suid mandible, showing broad, flat-bottomed grooves from the porcupine's incisors. *Bottom*, a partial bovid metapodial, showing both broad grooves and eroded-looking areas caused by licking and sucking the bone.

that animals going to drink might have stepped first on one end of the bone, driving it into the substrate and raising the other end into a nearly vertical position, and then on the upraised end, driving the entire bone into the mud. Such a sequence of events is a plausible, if unproven, explanation for the spatial distribution of these fossils.

TEMPORAL EVENTS

Changes in habitats, climate, or species abundance may cause differences in spatial distribution among different strata in the same locality. For example, deltas often build outward into large bodies of water, so that what was once a prodelta becomes a floodplain. Such a change will alter the proportions of different species represented in the fossil assemblage; aquatic species will decrease and, if there is a gallery forest along the river banks, the increasingly common terrestrial animals will be a mixture of forest- and savannah-dwelling species.

The clearest evidence of this type of change is sedimentological. Desert sediments differ from forest sediments, floodplain deposits differ from prodelta deposits, and so on. The most interesting case, however, is one in which the sedimentary environment, climatic regime, and agent of collection responsible for the assemblage remain constant, yet species abundance varies. In such instances, it can be safely concluded that evolutionary changes in the community have occurred, since taphonomic differences can be ruled out. In fact, the problem in trying to trace community evolution through time is that taphonomic differences can rarely be ruled out, because apparent shifts in faunal representation and inferred habitat may be caused by differences in the agent of collection and concentration responsible for the assemblage.

Summary

Spatial distribution data, though infrequently recorded by excavators in the past, can reveal a great deal about the paleoenvironment and the agents responsible for the formation of the assemblage. Patterning in the spatial distribution of bones, caused by such agents as geologic, biological, or temporal events, may be revealed by analyzing the distribution data by horizontal or vertical sections, or in terms of the orientation and dip of the long axes of the bones. All such data must be corrected, prior to analysis, for distortions produced by postdepositional dipping of the geologic beds.

Dip and orientation data are especially sensitive indicators of the effects, or their lack, of hydraulic forces. There are a number of theoretical and pragmatic problems in determining whether or not such data show a preferred orientation, and it is important to remember that intuitive evaluations of orientation data may be misleading. If biological events have produced patterning in spatial distribution data, the effects of all other known causes of patterning must be ruled out before it can be safely assumed that hominid activity was responsible for the observed patterns.

5 Tracing the Taphonomic History of an Assemblage

Although both the geological context in which fossils are found and their spatial distribution in the sediments may reveal considerable information about the paleoenvironment and site formation events, more detailed information about the taphonomic history of an assemblage must come from the bones themselves. The state of preservation of each fossil, as well as of the whole assemblage, reflects to a great extent the sequence of taphonomic events. For example, it may be possible to demonstrate that a bone was weathered, scavenged, and then buried in soils and infiltrated by the roots of plants.

By assessing the physical characteristics of an entire assemblage, one may be able to sort it into components with different taphonomic histories, which can be of great value in analyzing a mixed assemblage derived from different habitats. Analysis of the state of preservation may yield clues about how long the assemblage was exposed before being buried in sediments, and this information may alert the researcher to possible sources of distortion in the assemblage. For example, an assemblage that was exposed for a number of years before burial may lack all remains of smaller animals—juveniles of large species and adults of small species—because of their greater vulnerability to destruction by weathering and scavenging. In such an assemblage the researcher would look for individual bones showing tooth marks, punctures or depressed fractures from carnivore canines, or breakage characteristic of scavenging activities.

To decipher the physical evidence left by different taphonomic events, researchers must rely heavily on studies of what happens to modern bone assemblages in natural environments. Binford and Bertram (1977), Gifford (1977), Hill (1976, 1978), Behrensmeyer and Dechant-Boaz (1980), Shipman (1975), Shipman and Phillips-Conroy (1977), Payne (1965), Brain (1967, 1980), Sutcliffe (1970), Crader (1974), Haynes (1980), and Yellen (1977) are among those who have done such studies. Because of their work, it is increasingly possible to link the visible characteristics of fossils with particular events and even, in some instances, to estimate the duration of the events. Work now in progress using the scanning electron microscope (Shipman, in press) may yield a more precise definition of the changes produced in bones and teeth by different taphonomic events (see Chapter 8).

Six kinds of evidence are examined to establish the state of preservation of a bone or fossil.
1) Representation: completeness of the skeletal element; also, what portion of the skeletal element is preserved
2) Breakage: location, quantity, and type of breaks
3) Chewing or cutting marks: from the actions of predators, scavengers, rodents, other herbivores, insects, or hominids using tools
4) Abrasion: from the friction produced by colliding with other particles during hydraulic or aeolian transport
5) Weathering: from the effects of exposure to the elements
6) Strontium or other trace-element content: in bones or teeth as a result of diet

Each of these factors and its implications for reconstructing the taphonomic history of a bone, fossil, or assemblage will be discussed in turn. Many examples are taken from the analysis of the fossil assemblage at Fort Ternan (Shipman, 1977; Shipman, Walker, Van Couvering, Hooker, and Miller, 1981).

Representation

The representation of a skeletal element in an assemblage may vary from complete to fragmentary. Only an assemblage that is rapidly buried, probably one with many articulated skeletons, can be expected to have large numbers of complete bones. In most cases bones are broken by natural processes of decay, weathering, predator actions, hydraulic transport, trampling by herbivores, and diagenetic events (postdepositional geologic changes). However, many assemblages do include a high proportion of complete or nearly complete bones, perhaps because most of the assemblages that have suffered more violent taphonomic histories simply have not survived to become fossilized in noticeable concentrations.

In calculating the relative completeness of fossils, one needs as much information as possible about the techniques used to collect the assemblage. Many early fossil hunters were employed by museums interested in "good" or complete specimens for display; often they were also urged to bring back skulls or mandibles rather than other body parts. Such an assemblage will be strongly distorted in terms of both representation (completeness) and the abundance of different skeletal elements. Another issue is the ability of the excavators or collectors to recognize fragments of skeletal elements; many do not collect specimens judged to be unidentifiable scrap. However, others conscientiously preserve every scrap of bone or tooth from a site. Because collector bias can thoroughly distort an assemblage, there is no substitute for researching the methods used to collect the assemblage (see Chapter 7).

The skeletal elements present in an assemblage (skeletal abundance, or representation in another sense) may not include all the bones from the original animals (Figure 5.1). Teeth and alveolar regions of mandibles or maxillae may be found in far greater numbers than mandibular rami or vault bones, although each of these parts must have been present in the original animal. When such a disproportion is observed, the problem is to determine whether it has been caused by some biological agent of collection or by the differential durability and density of the skeletal elements. Disproportions in the representation of different portions of skeletal elements has also been observed, as when the proximal and distal ends of long bones are found in proportions considerably different from the 1:1 ratio in whole bones. Brain (1976) has shown that the proximal and distal ends have different amounts of spongy and compact bone (S/C ratio). Spongy bone, which is more fragile and

PRIMATE SKELETON

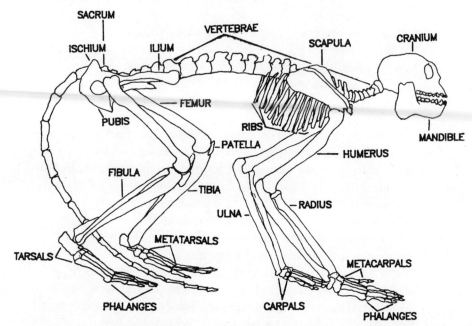

Figure 5.1 Skeletal elements of a typical primate and bovid. *Left*, a primate (*Mesopithecus*); and *right*, a bison. Not visible in these views are the sternum of each species and the clavicle of the primate. Note that the fused metatarsals of the bovid are called a cannon bone; sometimes this term is also used for fused metacarpals.

richer in hemopoietic tissues, is more attractive to predators and scavengers, so it is not surprising that scavenged assemblages include high proportions of those parts with less spongy bone. To determine whether the S/C ratio differed in proximal and distal humeri, Brain sawed off both ends of several humeri, sealed the exposed ends with a thin layer of wax, and submerged them in water to determine their specific gravity. The end with the higher specific gravity must contain more compact bone. It is important to note that the S/C ratio of proximal and distal ends varies according to skeletal element and species. However, this experiment is easily performed for any skeletal element or species.

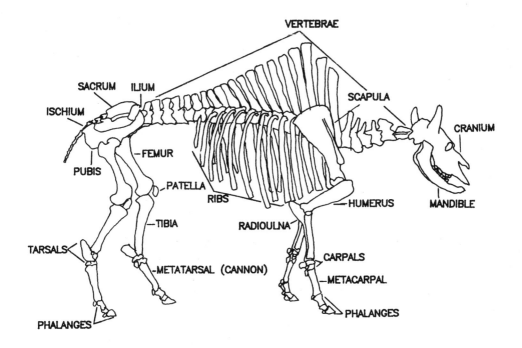

Binford and Bertram (1977) examined the differential survival of sheep bones exposed to scavenging by dogs. They found a distinct correlation among the density of the portion of each bone, the age of the animal, and the probability of survival of each portion of bone. They observed that for species that give birth seasonally, a graph of the density of parts of skeletal elements would show discrete clusters of data representing age classes. These data provide a useful analysis of the differential survival of skeletal elements exposed to medium-sized scavengers. It is unlikely that the survival probability of bones exposed to any taphonomic events would be identical to that documented by Binford and Bertram, but many of the same factors would pertain.

Breakage

The ratio of broken to unbroken bones in an assemblage may provide some measure of the destructive forces to which an assemblage was exposed prior to burial. However, breakage studies have a long and controversial history in paleontology. Dart (1949, 1957, 1959) has asserted that various types of breakage are distinctively hominid—most notably the spiral fracture, which he felt could be produced only by striking a long bone at midshaft and then twisting the ends in opposite directions. Although more recent studies have showed that this break may be produced by carnivores (Shipman and Phillips-Conroy, 1977) and trampling (Myers, Voorhies, and Corner, 1980), it is still assumed in many cases that spirally fractured bones were broken by hominids. Sadek-Kooros (1972) and Bonnichsen (1979) have experimented with certain techniques and methods of breaking bones. Particular kinds of breakage have also been taken as evidence of murder at australopithecine and other ancient sites (for example, Ardrey, 1961; also Brain, 1970; Roper, 1969), although increased appreciation of the many possible causes of breakage has reduced the popularity of such interpretations (see also Chapter 7, which discusses postfossilization breakage).

Problems in determining the agent of breakage have occurred repeatedly at various archaeological sites: animal bones or fossils from sites containing evidence of hominid technology (tools, hearths, dwellings) are usually taken to be those of prey species hunted and eaten by hominids. While it is unquestionably true that humans and their ancestors have long hunted and consumed animals, it is not true that broken animal bones or fossils in such situations always represent such events. Surveys of modern environments have revealed that ancient stone tools are often concentrated into lag deposits as wind or water erodes the land surface. The bones of any modern animals dying in or near these concentrations are not, of course, the prey of the hominids that made the tools. Some inferences can be drawn about specific taphonomic events from what little is known of the causes of different types of breakage (Figure 5.2). Green or fresh bones may produce sawtooth fractures, whereas older, dried bones more often show step fractures (Bonnichsen, 1979). Flaked fractures and de-

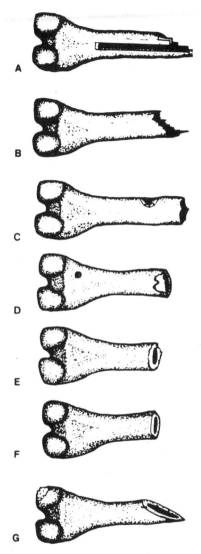

Figure 5.2 Breakage types, as shown on distal femora: *A*, a step or columnar fracture of the shaft; *B*, a sawtooth fracture of the shaft; *C*, a depressed fracture at the side of the shaft and a V-shaped break through the shaft; *D*, a puncture at the distal end and flaking (removal of the outer table of bone) more proximally; *E*, an irregular perpendicular break through the shaft; *F*, a perpendicular smooth fracture through the shaft (probably a postfossilization break); *G*, an irregular articular break on the condyle, exposing the cancellous bone of the interior, and a spiral fracture of the shaft. (From Shipman, Bosler, and Davis, 1981.)

pressed fractures often result from predator chewing but can be caused by hominids as well (Hill, 1975). As discussed earlier, Dart (1959) hypothesized that spiral fractures indicate hominid-induced breakage, but such fractures also occur in other situations producing torque.

To analyze breakage patterns and identify the most probable cause of breakage, the researcher must follow several simple conventions.

1) Compare like with like. First, compare skeletal elements that have similar basic structures. The structure of a skeletal element determines to a large extent how it will break, regardless of the cause. Long bones, which are basically cylindrical, are unlikely to show the same kinds of breakage as flat, broad bones such as scapulae or innominates. The differences in breakage are also a function of the microscopic structure of the bones, so breaks occurring in areas that are primarily spongy bone will differ from those occurring in areas of compact bone. Second, compare bones from similar animals. The size and structure of the animal have a marked influence on how its bones break. It is fruitless to look for similarities in the breakage of, say, elephants and ostriches, but two species of medium-sized bovids may usefully be compared.

2) Use standardized definitions of breakage types that can be recognized by different researchers. Following this convention ensures that analyses from different sites can be compared, and expected patterns of breakage can be compiled. Only when these patterns differ substantially from those known to be caused by other agents should hominids be considered the agents.

3) Consider the overall pattern rather than the individual bone. Very few types of breakage clearly identify the agent, one exception being the cutting marks made by hominids using tools. It is not usually possible to identify with certainty the agent of breakage of a particular bone, because many agents can produce similar types of breaks occasionally. However, overall breakage patterns, indicating the frequencies of different types of break, may point to the agent of breakage for a whole assemblage.

The first study to state and follow these conventions explicitly was that of Shipman, Bosler, and Davis (1981), which compared the breakage patterns in several Plio-Pleistocene assemblages from Africa. All skeletal elements from cercopithecids were examined; generally, breaks were compared only with breaks in the same body part. However, all metapodia were treated as a single group, as were all vertebrae, all ribs, and all phalanges. All available cercopithecid specimens from the Olorgesailie, Kenya, hand-axe site and from various living floors, or campsites, at Olduvai Gorge were examined. As a control, all cercopithecid material from all nonarchaeological sites at East Turkana, Kenya, was examined. Because East Turkana is broadly contemporaneous with Olduvai and Olorgesailie and because the sites share a similar fauna, it was assumed that all nonhominid causes of breakage (such as trampling, weathering, and carnivore activity) would also be observed in the East Turkana assemblage, and all deviation from the pattern found at East Turkana could be attributed to hominid activities. It was explicitly assumed that although some bones in the East Turkana sample may have been broken by hominids outside of recognizable sites, these would be insignificant compared to the large numbers of bones broken by other agents.

The following variables were recorded for each specimen: taxon; skeletal element; representation (proximal, distal, shaft, or other parts); number, location, and type of break; and maximum length of fragment. The classification of breakage types used in this study is shown in Figure 5.2.

Chi-squared tests demonstrated that the frequency of different types of breaks and the representation of parts of skeletal elements at Olorgesailie were significantly different from the frequencies at other sites. T-tests determined that there were significantly more breaks per mandible at both Olduvai and Olorgesailie and that the mandibular breaks occurred farther forward along the tooth row at these sites. But the overall pattern of breakage was most distinctive statistically at Olorgesailie, where several types of breaks at particular locations on particular skeletal elements were designated as probably hominid-caused. The breakage pattern of the Olorgesailie bones was as distinctive from that at Olduvai, the other hominid site, as from that at East Turkana, suggesting that the hominids at Olorgesailie were breaking these bones up in an unusual way. As a cautionary note, it should be noted that this study found significantly *lower* percentages of spiral fractures of humeri and femora at Olorgesailie, where the primates were judged to have been butchered, than at the other sites.

Chewing and Cutting Marks

Many modern species, including carnivores, rodents, and artiodactyls, chew bones regularly or occasionally. Porcupines and wolves also suck bones, leaving eroded areas. To judge from the marks found on many fossils, a number of ancient animals also chewed bones. In at least some cases, the chewing marks can be used to identify the species or types of species responsible. The marks range from small, flat-bottomed grooves resulting from incisal gnawing to large punctures or depressed fractures made by the canines of hyenas or other large carnivores.

Cutting marks can safely be assumed to be the work of hominids, using tools of stone, bone, shell, metal, or wood. Careful measurements may provide sufficient data to identify the type of tool used, by matching the angle of the cutting mark with the edge-angle of various tools (Walker and Long, 1977). Scanning electron microscope studies (Shipman, in press) reveal distinct differences between the marks made by stone tools and those made by carnivore teeth (Figure 5.3). Tool marks show many fine parallel striations within a main groove, presumably produced by irregularity of the cutting edge. Tooth marks usually show no fine marks within the main groove, but there are sometimes small perpendicular ridges produced by the "chattering" of the teeth across the surface as the animal applied force. In addition, punctures made by the canines of carnivores are different from the marks produced by striking a bone forcefully, perpendicular to the surface, with a stone tool (Figure 5.4).

The prevalence of chewing marks in an assemblage reflects both the length of time the bones were exposed prior to burial and the population density of bone-chewing species. Bones that are buried soon after the death of the animal have little chance to be chewed, whereas those exposed for weeks or even years in an area inhabited

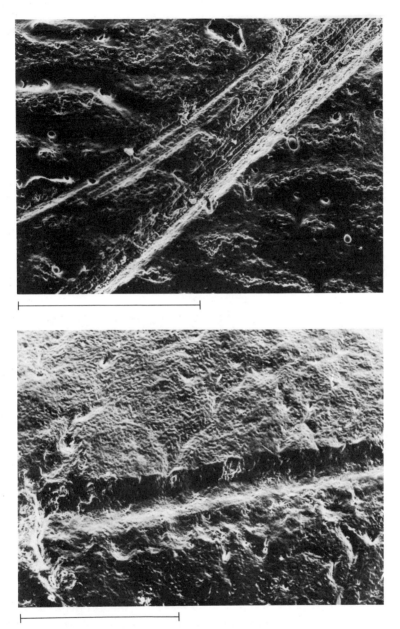

Figure 5.3 Top, scanning electron micrograph of a slicing mark made on fresh bone with a newly made stone tool. Note the fine striations within the main groove. The bar represents 1 mm. *Bottom,* scanning electron micrograph of a scratch made on a fresh bone by the teeth of a striped hyena. The bottom of the groove is smooth and rounded. The bar represents 1 mm. (From Shipman, in press.)

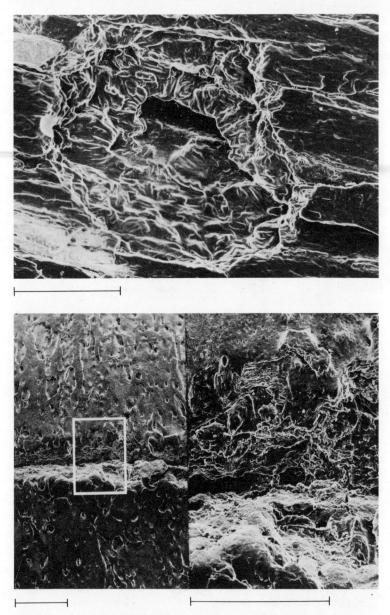

Figure 5.4 Top, scanning electron micrograph of a puncture caused by a carnivore's canine tooth. The bar represents 1 mm. *Bottom,* scanning electron micrograph of a chopping mark made by striking a fresh bone with a stone tool at a roughly perpendicular angle. Fragments of bone are crushed inward at the bottom of the groove. At the right is a close-up of the area in the box. The bars represent 1 mm each. (From Shipman, in press.)

by bone-chewing species are likely to show more extensive chewing marks. The season of the year when the animal died may also influence how fully a carcass is chewed and consumed, especially if the availability of prey varies seasonally (Haynes, in press; Kruuk, 1972). Gifford (1977) reported that chewed bones are rare among the modern bones she examined at East Turkana, which suggests low populations of both rodents and carnivores. This is confirmed by her observations of the local fauna. In other modern assemblages, such as that collected by Brain from a porcupine lair (1980), gnawing marks were found on more than 60 percent of the bones. Bearder (1977) reported that 94 percent of the bones from a spotted hyena lair were chewed; 77 percent chewed bones were found in an assemblage collected by striped hyenas (unpublished study by Shipman, Davis, and Bosler); Haynes (in press) reports gnawing of all skeletal elements retrieved from wolf dens, but very low percentages (5–17) of chewing or gnawing in assemblages from wolf kill sites. The most extreme example of bone-chewing is probably that reported by Kruuk (1972) in Ngorongoro Crater, Tanzania, where spotted hyenas chewed the skeletons of even large animals, such as rhinos, into unrecognizable fragments in a few days.

Carnivores seem to prefer particular skeletal elements, including the neural spines of vertebrae and the gonial angles of mandibles. They also frequently chew the ends of long bones, scapular blades, and ilia, presumably because of the hemopoietic tissues in these skeletal elements. Ribs may be cracked into small splinters, with only a few inches of each rib close to the vertebral column escaping destruction (Miller, 1969, 1975; Hill, 1980; Haynes, personal communication). The carpals and tarsals of primates may be completely devoured, but the same portions of bovids bear so little meat that carnivores often leave them intact (Brain, 1980). Potts (personal communication) has observed that hyenas often puncture the ends of metapodia with their canines, presumably to get at the marrow inside (Figure 5.5). Certain insects, such as termites, tanacid moths, and some beetle larvae, chew bones, leaving rounded bore holes or worm tracks (Figure 5.6). It is likely that ancient insects or insect larvae also chewed bones, because such marks are found on fossils.

Figure 5.5 Close-up of the ends of two radii of juvenile bovids, showing the punctures typically made by hyena teeth.

Figure 5.6 A scanning electron micrograph of a bore hole made by an insect that has burrowed into the bone. The bar represents 1 mm.

The presence of chewing marks on bones of different species can identify which species are most heavily preyed upon; this method was used on the Fort Ternan assemblage (Shipman, 1977; Shipman, Walker, Van Couvering, Hooker, and Miller, 1981). The five taxa that had chewing marks on 50 percent or more of the specimens examined included two species of giraffid, indeterminate giraffids, and two species of medium-sized bovids. It is highly probable that these species were favored prey. Rodents, carnivores, and proboscideans showed no signs of have been chewed.

Another approach to assessing the intensity of predator activity is to examine the frequency of chewing on a number of specimens of the same body part. When all body parts are inspected, a few individuals that have chewing marks on many skeletal elements may unduly inflate the frequency of chewing marks. With the single-element assessment, each specimen represents one individual, thus providing a minimum-frequency estimate; the other estimate is a maximum. Shipman (1977) has followed this method, inspecting right mandibles of a possible prey species from Fort Ternan because Hill's (1975, 1980) data showed that mandibles have chewing marks more frequently than many other body parts. Although inspection of all body parts showed that over 50 percent of the bones of this species were chewed, only 26 percent of the mandibles had been chewed; as predicted, the first method provided a maximum-frequency estimate.

Abrasion

Abrasion occurs because of the impact of wind- or waterborne particles on bone, or vice versa. Both aeolian and hydraulic transport produce abrasion and/or pitting of bones. In either case, edges of breaks and anatomical crests or ridges become rounded and are eventually obliterated. In addition, severe abrasion may remove the outer surface of bones, and Korth (1979) has observed that experimental abrasion of rodent mandibles often exposes the incisors. Regrettably little has been done to systematically analyze abrasion on fossils, in part because of the difficulties of assessing the degree of abrasion. The specimens must be inspected under a microscope, because detection of fine details is important. It is a task very similar to that of assessing the angularity or roundness of sedimentary particles, which is well known to be inexact (but see Krinsley and Donahue, 1968; Krinsley and Wellendorf, 1980). Another problem is the lack of experimental data for measuring the duration or velocity of transport that produces abrasion. It is possible to say only that unabraded bones have not been transported any substantial distance and that heavily abraded bones have been transported over considerable distances.

Because of the difficulty in distinguishing among finely graded degrees of abrasion, to date these three broad classes have been used: (1) little or no abrasion—fresh, sharp edges or breaks; (2) moderate abrasion—some rounding of edges or breaks; and (3) heavy abrasion—edges obscured, breaks well rounded, surface bone possibly missing (Shipman, 1977); see Figure 5.7. The breadth of these categories reflects the lack of knowledge of the circumstances that produce different degrees of abrasion. However, given the modest amount of available information, a surprising amount can be learned from abrasion data, even with these broad and imprecise classes. Marked differences in abrasion on the bones

Figure 5.7 Categories of abrasion. *Left to right:* unabraded, moderately abraded, and heavily abraded bones. The moderately abraded bones show some rounding, but many fragile flakes still remain.

of different species can be used to separate a mixed assemblage into its components when there are other data (weathering, skeletal representation and abundance, known habitat preferences) available to help define the various original communities (see Shipman, 1977).

Weathering

Weathering is the damage to body parts produced by exposure to the elements prior to deposition. Bones and teeth may be heated, cooled, wetted, dried, frozen, and thawed; these processes slowly break down the skeletal elements and will eventually destroy them if they are not protected by burial in sediments. Postdepositional weathering and erosion do, of course, occur, but are of little use in reconstructing paleoenvironments or animal communities. Behrensmeyer (1978) has developed a sequence of weathering stages that seems applicable to bones in many environments. These stages were used in the study of modern bones in Amboseli and in Gifford's study of modern bones at East Turkana (1977) and were confirmed by Shipman's study (1977) of the weathering stages of Fort Ternan fossils (see Figure 5.8). A low-powered microscope is useful in determining the depth and extent of cracking and exfoliation in order to identify the weathering stage of a specimen. Even so, it can be extremely difficult to distinguish between severely weathered and severely abraded bones.

Although the severity of weathering is directly related to the length of exposure of the bones, weathering proceeds at different rates in different microenvironments. Thus, for example, Gifford (1977) observed a difference in the weathering of a hippo tibia and a radioulna from the same individual when one was lying in a slightly damper place than the other. Similar observations have been reported by several researchers (see also Brain, 1967; Toots, 1965b; Behrensmeyer, 1978; Haynes, personal communication).

The influence of microenvironments makes it more difficult to determine a standard rate at which bones pass through these weathering stages. Gifford's data, although they may not be applicable to other environments, show that bones generally progress through stages 1 and 2 within the first two years. Weathering then

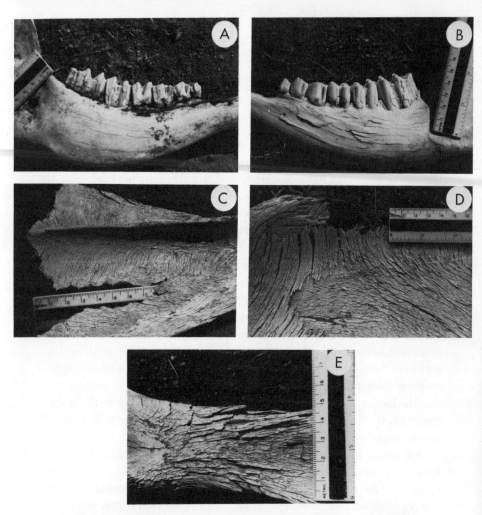

Figure 5.8 Weathering stages on modern bones. *A*, stage 1: superficial longitudinal cracking; *B*, stage 2: flaking of outer surface; cracks, 1–1.5 mm deep; *C*, stage 3: patches of fibrous bone; *D*, stage 4: deeper cracking and extensive flaking; *E*, stage 5: bone falling apart *in situ*. (From Behrensmeyer, 1978; reprinted by permission of the author and *Paleobiology*.)

seems to slow down, and three years after death many bones are still at stage 2 or 3. Gifford suggested that it may take as much as ten to fifteen years for bones to reach stage 5 in a semidesert environment like that of East Turkana. Partly buried or subsurface bones may take even longer to weather to stage 5 (Table 5.1). Gif-

Table 5.1 Weathering stages of surface and subsurface bone samples from seven age-ranked sites in the Il-Erriet Valley, Ileret, Kenya.

Site number	Age of bones[a] (years)	Weathering stage		
		Rib surface/subsurface	Long bone surface/subsurface	Dense bone surface/subsurface
301	11–15	4/1	4/1	3/0
310	9–13	3/2	3/2	3/0
308	10–13	4/1	3/0	2/0
306	6–10	3/1	3/0	3/0
302	4–5	4/1	3/1	3/0
304	4–5	3/0	3/0	2/—
307	4	3/0	2/0	2/0
311	1–2	1/0	1/0	—/—

Source: Gifford (1980). (From *Fossils in the Making*, A. K. Behrensmeyer and A. P. Hill, eds., reprinted by permission of the authors, editors, and the University of Chicago Press © 1980, The University of Chicago.)

a. The number of years site was occupied before samples were collected in 1974 is used as the maximum age of the bones.

ford also found that equid bones weathered more slowly and were more resistant than bovid bones, probably as a result of their greater robustness and density. Haynes (personal communication) has documented roughly similar rates of weathering in Canada.

In her initial survey, Gifford did not observe any bones at stage 5, perhaps because bones at this stage are nearly unrecognizable and almost always unidentifiable. Shipman (1977) also failed to find any stage 5 fossils in the Fort Ternan assemblage, which might be attributed to the same problem of recognition. However, on the whole the Fort Ternan assemblage is so fresh that perhaps no bones weathered long enough to reach stage 5.

A few important patterns have been noted about weathered bones. The different stages are distinguished in part by the presence and the severity of fine cracks, which may develop quite rapidly in dry, hot environments. The cracks are almost invariably parallel to the predominant alignment of collagen fibers. The cracks deepen, and eventually a piece of bone may be detached; commonly, the base of the mandible splits off from the rest of the corpus, or the lateral and medial sides of metapodia separate as a result of longitudinal cracking (Figure 5.9). Before studies of

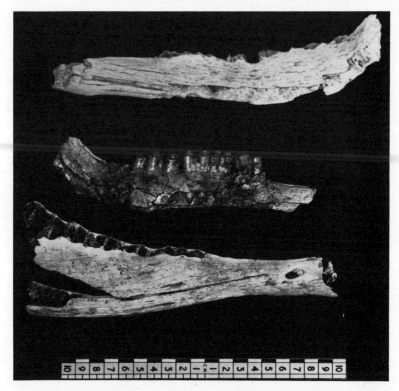

Figure 5.9 Top to bottom: a modern antelope mandible with the bottom of the corpus missing; a fossil antelope mandible with an identical break; a modern antelope mandible showing that such breaks occur by the propagation of longitudinal cracks through weathering.

weathering were made, both these effects had been attributed to hominid activities. Of course, as Noe-Nygaard (1977) observed, hominids do break off the bases of mandibles.

Bonnichsen (1979), examining fossilized bones with a scanning electron microscope, observed networks of microscopic cracks that probably developed during prefossilization drying and weathering. He believes that these cracks may guide the development of large-scale fractures when fossils are broken and may thus cause the distinctive shapes of fragments of bones broken after fossilization. Weathering of long bones, such as tibiae or humeri, that are normally subjected during life to torsional stresses, can result in spiral fractures similar to those resulting from hominid activities (Sadek-Kooros, 1972; Noe-Nygaard, 1977; Hill, 1976; Myers, Voorhies, and Corner, 1980).

The following observations have been made about the weathering of different skeletal elements from a single individual, regardless of species (unpublished study by Shipman, Davis, and Bosler). First, skeletal elements of a single skeleton, whether modern or fossilized, commonly exhibit more than one weathering stage. When partial or complete burial of some elements has occurred, a range from stage 0 to stage 2 has been observed on modern bones. Second, as noted by Behrensmeyer (1978), carpals and tarsals are frequently less weathered by one or more stages than other bones of the same individual. These are often among the first bones to become buried by trampling as well. Third, animals seem to avoid stepping on skulls, which are often kicked or otherwise displaced from the rest of the skeletal elements (Haynes reports similar observations for caribou and bison skulls). Therefore, if trampling has been a major factor in burial, fossil skulls may exhibit a more advanced weathering stage than the rest of the skeleton.

Strontium and other Trace-Element Content

Toots and Voorhies (1965) pioneered the efforts to measure the mineral content of fossils as an indicator of diet. They found that the bones of animals in the major dietary groups (browsers, grazers, and carnivores) showed different ratios of strontium to calcium. Browsers ingest more strontium from leaves than grazers ingest from grasses, and carnivores were found to have the lowest strontium levels, because animal flesh contains relatively little of that element. Carnivores that eat bones as well as flesh ingest more strontium than flesh-eaters but less than herbivores.

There are several important prerequisites to a successful strontium analysis. The strontium level in bones, both fossilized and modern, is not physiologically regulated and is not subject to diagenetic changes (Parker and Toots, 1980); therefore it reflects the animal's ingestion of that element. The primary source of strontium for land vertebrates is food, which in turn is derived from the local environment. For aquatic vertebrates the primary source is water (Rosenthal, 1963). Strontium levels in modern herbivores from some parts of Nebraska are reported to be three to four times greater than in herbivores from other parts of the state (Parker and Toots, 1980). Therefore it is very important that the fossil assemblage under analysis be unmixed, so that differences in environmental levels of strontium do not produce confusing and inaccurate conclusions about dietary habits.

A strontium analysis of fossils from the Shungura Formation in the Omo, Ethiopia (Boaz and Hampel, 1978), reveals some of the difficulties of the technique. Most of the fossiliferous localities in this formation are the result of fluvial deposition; the geologic contexts range from low-energy environments to channel and point bar deposits. Thus, although the assemblages were excavated from the same localities, they can be presumed to be mixed to varying degrees and therefore may not necessarily be reliable subjects for strontium analysis. The results of this study were confusing. For example, in one locality leopards (flesh-eaters) showed a strontium content higher than colobines (browsers); further, some species showed tremendous variability in strontium levels from sample to sample yet probably did not vary their diet widely. Boaz and Hampel concluded that strontium is probably not preserved in the Omo fossils in the amounts that were present in life. They suggested that these fossils, at least, were altered by the strontium levels in soils and groundwaters, but this explanation does not explain directly the unusual variability they found. Their problematical results underscore the need to select samples for strontium analysis with great care, for only unmixed and untransported assemblages are likely to give meaningful results. Further, more experimental work is needed to resolve the question of whether or not there is postdepositional alteration of the strontium level or of the strontium-calcium ratio.

Analysis of State of Preservation Data

What can be learned from studying the state of preservation of an assemblage? In this chapter different measures of the state of preservation have been presented. Some of these variables are less informative than others because at present there is little experimental data on the effects of different events on bone preservation. Other may seem redundant; for example, if the sedimentary evidence suggests a fluvial environment, is it necessary to examine hundreds of bones under a microscope in search of abrasion? The answer is yes, for a variety of reasons. First, studies now in progress will yield much more information that is needed for interpretation of preservation data. Second, even if only the most tentative deductions can be drawn from such data, it is still important to document, compare, and contrast the state of preservation of fossils from different assemblages. This process will increase our knowledge of the range of states of preservation that characterize fossils from different environments. Establishing differences or similarities in the state of preservation of channel assemblages, for example, may prove important in refining our understanding of how bones are collected and deposited by water currents. Finally, although each type of data may contribute but a small amount of information about the past, by synthesizing many diverse types of data we can make a detailed composite reconstruction. The level of resolution is always gross in paleoecology and taphonomy, and every piece of evidence, however small, is important.

Synthesizing the taphonomic data from a site or assemblage may suggest scenarios to account for different evidence. Sometimes the reconstructions must be less detailed, as when the evidence of abrasion, geology, and skeletal representation indicates that the fossil assemblage has been hydrodynamically winnowed. In such cases the proportions and diversity of species in the fauna are likely to be poor reflections of the original animal community. Or two different components may be distinguished in an assemblage, one believed to be locally derived and the other transported from some distance. Unfortunately, at present we may not be able to determine the distance the bones were transported or the habitat from which they were derived. At best, data on the state of preservation of bones can be integrated into a detailed and useful reconstruction of the original habitat and the surrounding animal communities; at worst, such data can serve as a warning about possible sources of distortion in the assemblage as it is found.

6 Faunal Analysis

Faunal analysis is a technique commonly employed in reconstructing both the local paleoenvironment and the taphonomic history of an assemblage. Traditionally, the objective was to compile a list of the species found at a particular site. However, this is a trivial piece of information, of value only for superficial or preliminary comparison between sites. Increasingly, paleontologists seek to determine not only the species represented at a site, but also their relative abundance, the representation of skeletal elements, and their population (age) structure. From this information researchers can derive an approximation of the original animal communities as well as evidence of the taphonomic history of the assemblage or its components. By plotting the rise and fall of species' abundance and the development of adaptations and specializations, it is possible to trace the interactions among species over time. Any evidence of hominid activities is important, for careful work may reveal what animals the hominids lived with or ate. Variation in the fauna over time is also important, for it may reflect habitat changes, evolutionary trends, or changes in hominid behavior. From faunal analyses come some of the most interesting details about the past, those that permit reconstruction of ancient ways of life.

In an assemblage from an archaeological site, the fauna is often assumed to be prey species or food in the diet of hominids. Sometimes the numbers of different species found at a site are used to extrapolate the size of the hominid population that could have been supported by such prey. This technique relies on standard estimates of the available meat per adult animal of each species (for example, White, 1952, 1953, 1954, 1955). Unfortunately, this procedure is vulnerable to three major sources of error.

First, the association between the bones of animals and the evidence of hominid presence (artifacts, hearths, potholes, or other structures) may be entirely fortuitous (Figure 6.1). This source of error is often dismissed as trivial, but it is important to seek evidence that the fossilized bones are clearly related to the hominid activities. Direct evidence could be cutting marks on the bones, which can sometimes be linked to the type of tool that produced them (Walker and Long, 1977). Another type of direct evidence would be the deliberate alteration of bones for use as tools or weapons, although the deliberateness may be difficult to demonstrate (Bonnichsen, 1979; Dart, 1959). If a site was occupied over a period of months or years, the researcher should find considerable variation in weathering of bones from prey species; conversely, if a site represents a single hunting or butchering event, the weathering should be similar on all bones in the assemblage. Of course, at early sites where the hominids' technological capabilities were limited, it is especially difficult to obtain direct evidence that bones represent prey remains. In such instances, examination of various aspects of the faunal representation and the state of preservation of the bones (including breakage patterns) is necessary to rule out nonhominid causes of bone alteration and prey selection.

The second source of error lies in the estimation of available meat on an adult of a given species. The most commonly used figures are those of White (1952, 1953, 1954, 1955), which later workers have suggested are unrealistically high (Lyman, 1979), because some were calculated from well-fed domesticated animals. Even if accurate estimates are obtained, the chances of error increase dramatically as the age of the site increases. Differences in weight between modern and ancient representatives of a species or genus are unmeasurable, and it is even more difficult to estimate the available meat of a now-extinct species. However, approximate values can be reached by careful analysis of osteological materials,

Figure 6.1 The humerus of a modern dog found partially buried and associated with two prehistoric microliths (marked by arrows). The manufacturers of the stone tools lived thousands of years earlier than the dog, yet if this were a fossil site, they would appear contemporaneous. This is an instance of accidental association of stone artifacts and bones.

and when used with appropriate caution and applied only to adult individuals, such estimates may be of value.

The third problem with such analyses is calculating the size of the hominid population that might have been supported by a given supply of meat. Modern human populations vary widely in both the importance of meat in their diet and in the "normal" caloric and protein intake (Cannon-Bonventre et al., 1977). Because vegetable remains are extremely prone to decay, it is exceptionally difficult to estimate the quantity and identity of nonmeat resources exploited by a population. As has been shown by Coe and Flannery (1964), there may be a substantial difference between the resources available to a population and those actually utilized. There is at present no means of reducing or even assessing errors in measuring variability in diet.

Identification

The first step in faunal analysis is identification of the specimens in an assemblage, which must be done by a skilled researcher familiar with the species or genera likely to occur. Easy access to a comparative collection of fossil and modern species is also crucial, although this is often difficult in the field. However, the importance of verifying and refining field identifications cannot be overemphasized; even a highly trained faunal analyst needs to refer to comparative materials frequently. More important, all subsequent stages of analysis are wasted if the identifications are inaccurate.

A specimen can be identified at one or both of two levels, skeletal element and taxon. If enough of a bone is present, it can be identified as to skeletal element. Identification of the taxon requires recognizable taxon-specific characteristics of shape, size, structure, or a constellation of these. As a general rule, if a specimen is so badly broken or damaged that it is unidentifiable at the first level, it will be unidentifiable at the second as well. A few taxa have such distinctive bone structure that the taxon can be identified even without identification of the skeletal element, but usually such taxonomic identifications are so broad—reptile versus mammal, for example—that they are of little use.

The factors that allow identification of a particular specimen at either level are first, the distinctiveness of the skeletal element in terms of shape, markings, or texture; and second, the likelihood of preservation, or preservation potential, of that element. The first factor is fairly obvious. A skeletal element with a distinctive shape or markings or texture—such as teeth or vertebrae—may be identifiable from even small or badly damaged fragments. Some taxa have skeletal elements that are so distinctive as to be identifiable at both levels from even a small fragment: examples are ostrich eggshell, aardvark teeth, or the ornamented and expanded ribs of the Miocene rodent *Afrocricetodon songhori* (Figure 6.2). In order for a distinctive skeletal element (DSE) to be identified as to taxon as well, it must possess some taxon-specific characteristics.

The second factor in identification involves the inverse relationship between the degree of damage to a bone and its preservation potential. Bones that are likely to be preserved, either because of their structural qualities or because of their placement within the

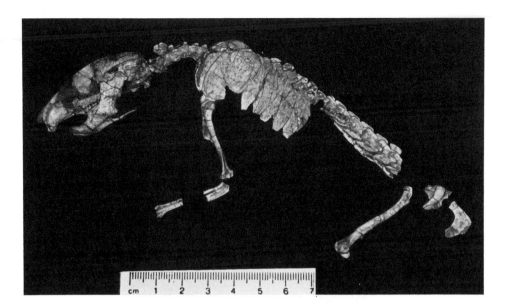

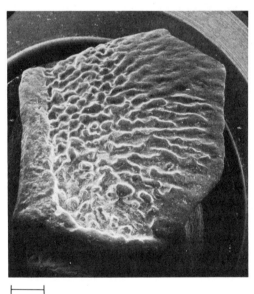

Figure 6.2 Top, nearly complete skeleton of *Afrocricetodon songhori* has peculiar expanded and ornamented ribs unlike those of any living animal. *Bottom*, a close-up of the ornamentation on a right rib; anterior is to the left in this scanning electron micrograph. The bar represents 1 mm.

body, are more likely to retain their identifying characteristics. This means that a DSE with a low preservation potential will quite often be identifiable at the first level but rarely at the second. This pattern applies to ribs and vertebrae, which share a complex of traits: low bone density (or high S/C ratio), high SA/V ratio, high attractiveness to scavengers, and lack of taxon-specific features over much of their surface. Generally speaking, once such bones are exposed to any destructive forces, they become very difficult to identify as to taxon. The skeletal elements that are likely to be identified at both levels also share a complex of traits related to high preservation potential and durability. Teeth (and body parts bearing teeth) are the preeminent example, but the larger long bones such as femora and humeri are also readily identifiable in many cases.

This inherent difference in the identifiability of different skeletal elements has produced a tremendous bias in the specimens collected. Teeth, skulls, and mandibles are usually collected disproportionately to their true representation in an assemblage, whereas postcrania—especially ribs, vertebrae, and fragments of long bone shafts—may be ignored because of the greater difficulty in identifying them at the second level. Of course, what is collected may reflect the training and capability of the collector as much as the differences in identifiability of the skeletal elements (Gray, 1979); see also Chapter 8.

Nearly all of the literature and most collectors fail to appreciate another source of paleoecological and taphonomic information in any fossil assemblage: the indeterminate fragments. These are often left on the site with the rubble or are retained only if there is a need to search for broken fragments of important specimens. But there is much information to be gleaned from analyzing the indeterminate fragments, which include specimens of unknown taxon (anonymous bones) as well as specimens that cannot be assigned to any skeletal element (unidentifiable bones). As a class, indeterminate fragments have several characteristics in common. By definition they are pieces of bones, and they have been subject to more destructive forces than identifiable bones or have been more affected by those forces, which have removed the characteristics that render whole bones identifiable.

Johnson (1960), dealing with shells and marine environments, projected three models of environment and burial circumstances that result in different proportions of fragmented shells in an assemblage. Although shells and bones are not equally vulnerable to breakage, the concept that differences in taphonomic history will be reflected in the percentage of fragments found in an assemblage is nonetheless applicable to fossil bones.

The indeterminate component of an assemblage can be analyzed in several ways. First, the indeterminate fragments can be compared with the identifiable bones to gain particular sorts of taphonomic data. If the two components consist of bones that are similar in size, degree of abrasion, and degree of weathering and gnawing, then probably the indeterminate fragments had the same taphonomic history as the others. In this case, they represent the most fragile and vulnerable bones of the original assemblage. Differences between the indeterminate and identifiable components in size, abrasion, or weathering would be evidence of different taphonomic histories; the indeterminate bones may be derived from a different animal community and probably were exposed to different destructive forces before burial.

In particular, if the indeterminate and the identifiable bones are of about equal size, then weathering, transport, and decay are likely to have been the destructive forces. In that case, the indeterminates have been so badly damaged that any identifying features are gone. On the other hand, if the indeterminate bones are smaller than the identifiable bones, breakage by predators, scavengers, or diagenetic events is likely. The indeterminate bones have been broken into pieces that do not retain the characteristic features or shape; they are probably pieces of bones that are especially attractive to scavengers because of the soft tissues or marrow associated with them. Finally, if the indeterminate bones are larger than the identifiable bones, the identifiable bones may have been consumed, concentrated, and transported in the digestive tracts of predators and thus protected from the forces that damaged the indeterminates.

A second type of analysis is comparison of the percentage of indeterminate bones in one assemblage with that in other assemblages as a gross indication of the effectiveness of the destructive forces in their taphonomic histories. A low percentage of indeterminate bones can be taken to mean that destructive forces were ineffectual and that the original death assemblage has probably been fairly completely preserved. A high percentage of indeterminates shows that strong destructive forces were at work, and the bones most vulnerable to those forces are likely to be absent or in a particularly poor state of preservation.

Percentages of indeterminate fragments in excavated assemblages have rarely been recorded, two exceptions being sites at Olduvai Gorge and Fort Ternan. At various Olduvai sites, Leakey (1971) reported that the indeterminate fragments made up from 3.6 to 68.9 percent of the faunal assemblage. There was no apparent correlation between the percentage of indeterminate fragments and the nature of the site. Assemblages from sites thought to be living floors or occupation sites included 5.4 to 68.9 percent indeterminate fragments, spanning nearly the entire range of variability at Olduvai. Other types of sites, such as those characterized by Leakey as having materials scattered throughout a considerable thickness of fine-grained tuff or clay, showed a similar range of variability. Sieving the sediments at one site, the MNK Skull site, raised the proportion of indeterminate fragments from 57 percent to nearly 98 percent. As Leakey pointed out, sieving at other sites would very probably produce similarly dramatic increases in the numbers of indeterminate fragments.

At Fort Ternan all rubble from FT-B was collected and sieved, but only 70 percent of the bones were of indeterminate species, and 54 percent of indeterminate skeletal element (Shipman, 1977). Thus the numbers of both anonymous and unidentifiable bones are comparable to those at the Olduvai sites that had *not* been sieved. If both sites had been sieved, the actual proportion of indeterminate fragments would probably be much higher at Olduvai than at Fort Ternan. The sites are quite different from one another in other ways. The Fort Ternan site was not a living floor of *Ramapithecus* in any sense comparable to that in which some of the Olduvai sites were living floors of *Australopithecus*. Also it is highly improbable that *Ramapithecus* was responsible for the bone breakage at Fort Ternan (*contra* Leakey, 1968; Ardrey, 1976). If it is postulated that sieving the living floors at Olduvai would markedly increase the numbers of indeterminate fragments, then it may be hypothesized that hominid living floors show a higher proportion of badly broken bones than do other types of sites. However, to support that hypothesis it would be necessary to sieve all sites at Olduvai and find that the proportions of indeterminate fragments increased markedly only at the living floors. It is hoped that the environmental and behavioral correlates of the percentages of indeterminate fragments in assemblages will become clearer as more work is done.

Abundance of Skeletal Elements

Once the specimens have been identified as to skeletal element, it is possible to establish the relative abundance of different skeletal elements, or the skeletal representation. This may help identify the agent of concentration, many of which act differently on different body parts.

INFERENCES BASED ON SKELETAL REPRESENTATION

Evidence for hydrodynamic sorting may be obtained from the data on skeletal representation in an assemblage. Voorhies' work (1969) and its extensions by Behrensmeyer (1975) and Korth (1979) have shown that currents usually sort the skeletal elements of mammals into three groups (see Chapter 2). If an assemblage contains skeletal elements from only one of these Voorhies Groups, water-sorting is indicated, and it may even be possible to establish the probable velocity of the current. On the other hand, an abundance of elements from all three Voorhies Groups strongly suggests that water currents were not an important force in concentrating the bones. More generally, currents may sort bones according to the size of the animal, so that assemblages containing the remains of only very large or very small animals may also be water-sorted. Sedimentological evidence is of great importance in assessing the agent of concentration. It is unlikely that an assemblage in a bed showing no evidence of hydraulic activity has been water-sorted, although waterlaid beds may contain unsorted assemblages. It is also important to ascertain whether the bones are associated with sedimentary particles having similar hydrodynamic properties.

Predator-collected bone assemblages also show a typical profile of skeletal representation: long bones, especially metapodia, are very common, as are mandibles and maxillae; vertebrae and ribs are poorly represented. In addition, the bones are likely to show tooth marks and patterns of breakage characteristic of particular carnivores. Hill (1980) and Haynes (1980) have investigated the patterns of breakage characteristic of several modern carnivore species. It has been observed (Brain, 1980) that carnivores often preferentially collect the bones of other carnivores, so that their representation in the resultant assemblages exceeds that predicted by normal predator/prey ratios. In an assemblage analyzed by Mills and Mills (1977), nearly 31 percent of the skeletal remains in a brown hyena lair were those of carnivores. Of course, the older the site, the less reliance can be placed on predictions based on the behavior of modern animals.

Predatory birds, such as owls, vultures, eagles, and other raptors, produce dense concentrations of small mammal bones by regurgitating bones, fur, and other indigestible material in pellets, often at their roosting sites (Mellett, 1974; de Graaf, 1960; Davis, 1959; Korth, 1979; Dodson and Wexlar, 1979). One of the South African australopithecine caves, Swartkrans, contains a rich deposit identified as being of owl pellet origin. Six characteristics can be used to identify such assemblages:

1) Only small animals, predominantly rodents, are represented.
2) Skulls and mandibles are often left intact. If the skull is damaged, the occipital region is broken or missing.
3) Hindlimbs are represented less frequently than forelimbs.
4) Bones do not necessarily show a high rate of breakage, depending on the raptor species; some articulated specimens are found.
5) Some bones, especially mandibles and pelves, show solution effects, such as feathering and thinning at breaks or margins.
6) A high percentage of skeletal elements is represented.

Modern assemblages collected by harvester ants are similar to owl pellet assemblages in many ways (Shipman and Walker, 1980). Important differences are that harvester ant assemblages, if fossilized, are likely to occur in paleosols that show evidence of the ants' tunnels; that the robust bones of the limbs and the mandible are most commonly preserved; that skulls are rarely intact but may be represented by unbroken nasals, frontals, occipitals, or parietals; and that the rodents and other small species represented show a wider range of habitats and niches than those collected by avian or mammalian predators.

Porcupines, although not predators, are also active bone collectors; apparently, they like to gnaw on hard objects with their prominent, ever-growing incisors. Brain (1980) observed a collecting rate of thirty-eight bones or other hard objects per lair per year in an area where there was an abundance of bones available. The modern assemblage in Nossob, South Africa, that he studied displayed a number of distinctive characteristics that would enable identification of porcupine-collected assemblages in the fossil record. These characteristics include a scarcity of fresh bones (less than 1 percent appeared to be fresh, which would be evidenced as a predominance of weathered bones in the fossil record) and indications of gnawing on a high percentage (60–69 percent) of the bones.

Brain also found that the skeletal representation in an assemblage from a porcupine lair contrasted sharply with that from a heavily scavenged assemblage of goat bones discarded by Hottentot villagers (Figure 6.3). The porcupine assemblage showed a high percentage of many skeletal elements that do not survive well under most conditions: vertebrae (including sacral vertebrae), ribs, and pelves. This was confirmed by calculating the survival rate or percentage representation (number present divided by the number expected) of the skeletal elements (Table 6.1). In further contrast to some carnivore-collected assemblages, less than 1 percent of the Nossob assemblage was carnivore bones.

In the same study, Brain documented the maximum length and weight of the bones collected by porcupines, which can be compared with unpublished data collected by Shipman, Davis, and Bosler on bones from a striped hyena lair from Kenya (Figure 6.4). The mean lengths of the bones in the two assemblages are comparable, but those collected by porcupines show a greater range of lengths (about 0.5–36 inches) than those collected by hyenas (0.5–9.5 inches). Similarly, the mean weights are comparable, but the porcupine assemblage covered a greater range of weights (the

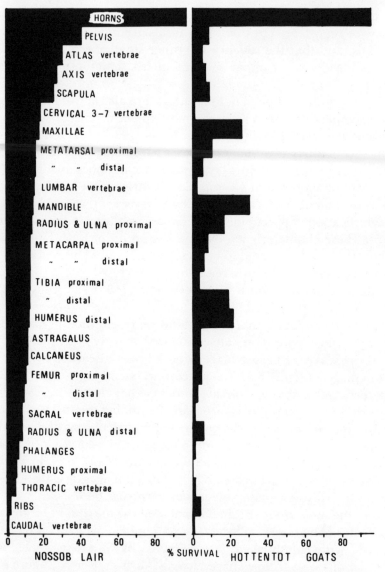

Figure 6.3 Bar graphs comparing the representation of various body parts in the Nossob porcupine lair and in a heavily scavenged assemblage of goat bones from a Hottentot village. (From Brain, 1980. Reprinted from *Fossils in the Making*, by A. K. Behrensmeyer and A. P. Hill, eds., by permission of the author, the editors, and the University of Chicago Press. © 1980 by the University of Chicago.)

Table 6.1 Percentage survival figures for bovid skeletal parts in the Nossob porcupine collection compared with a Hottentot goat bone sample.

Part	Nossob sample: 81 bovid individuals			Hottentot goat bone sample percent survival
	Number found	Original number	Percent survival	
Horn pieces	157	162	96.9	94.7
Pelvic pieces	66	162	40.7	9.0
Atlas vertebrae	25	81	30.9	6.3
Axis vertebrae	23	81	28.4	7.4
Scapula pieces	42	162	25.9	9.2
Cervical vertebrae, 3–7	76	405	18.8	1.2
Maxillae	15	81	18.5	26.3
Metatarsal, proximal	26	162	16.0	10.3
Metatarsal, distal	26	162	16.0	5.3
Lumbar vertebrae	76	486	15.6	2.7
Half mandibles	24	162	14.8	30.7
Radius and ulna, proximal	23	162	14.2	17.1
Metacarpal, proximal	22	162	13.6	8.4
Metacarpal, distal	22	162	13.6	6.0
Tibia, proximal	21	162	13.0	3.4
Tibia, distal	21	162	13.0	19.0
Humerus, distal	20	162	12.3	21.5
Astragalus	19	162	11.7	4.2
Calcaneus	18	162	11.1	3.7
Femur, proximal	17	162	10.4	4.7
Femur, distal	15	162	9.3	2.4
Sacral vertebrae	7	81	8.6	0.5
Radius and ulna, distal	14	162	8.6	5.8
Phalanges	60	972	6.2	0.9
Humerus, proximal	9	162	5.5	0
Thoracic vertebrae	53	1,053	5.0	0.9
Ribs	43	2,106	2.0	3.4
Caudal vertebrae	1	810	0.1	0

Source: Brain (1980). (From *Fossils in the Making*, A. K. Behrensmeyer and A. P. Hill, eds., reprinted by permission of the authors, editors, and the University of Chicago Press © 1980, The University of Chicago.)

smallest bone was between 0 and 50 grams, the largest 750 grams) than did the hyenas (range: 0.5–198 grams). Since porcupines weigh slightly less than striped hyenas, presumably hyenas are capable of carrying bones that are as heavy as those carried by the porcupines.

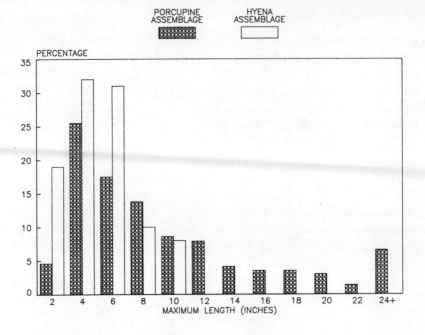

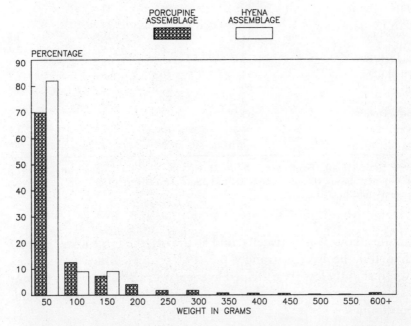

Figure 6.4 *Top*, lengths of bones collected by porcupines compared with those collected by hyenas. *Bottom*, weights of bones collected by porcupines and by striped hyenas. (Porcupine data from Brain, 1980; hyena data from an unpublished study by Shipman, Davis, and Bosler.)

Most probably the small size of the bones collected by the striped hyenas in this study reflects their domination by the sympatric spotted hyena. Work by Kruuk (1976) and Skinner et al. (1979) suggests that striped hyenas are more aggressive in hunting and social behavior in the absence of spotted hyenas; striped hyenas in India and the Near East probably collect larger and heavier bones than they do in Africa, where the spotted hyena is common. Although differences in the degree and type of mineralization at different sites mean that the comparison of weight of fossils from different sites is a spurious exercise, it is possible to estimate the prefossilization weight of a bone. In addition, calculation of the percentage of gnawed bones in each weight and length class shows that porcupines collect many small bones but preferentially gnaw the heavier and larger ones. Brain related this observation to the difficulty porcupines might experience in holding smaller bones between their paws while gnawing. Striped hyenas also tend to leave more tooth marks and punctures on the larger bones, but most of the smaller bones are fragments broken off the larger bones and therefore may be less likely to preserve gnawing marks. Thus the lengths and weights of bones and the patterns of damage can be used to distinguish between these agents of collection.

MATHEMATICAL ASSESSMENT OF SKELETAL ELEMENT ABUNDANCE

Comparing the representation of different body parts in different assemblages is a major problem. Equally problematical is assessing the significance of observed similarities or differences in skeletal representation. Visual comparisons, as in Figure 6.3, are striking but difficult to evaluate. Correlation coefficients (Pearson's r) have been used to compare the skeletal representation in different assemblages, but the results are meaningful only under the following circumstances:
1) The two assemblages are of similar size
2) Only skeletal elements present in both assemblages are considered
3) The abundance of the different elements approximates a normal distribution in each assemblage
4) The range of species is roughly comparable in each assemblage

Comparisons of data on one type of animal, such as bovids or primates, are especially informative.

If the criteria listed above are not met, the results of the calculation of r will be misleading in some way. A correlation coefficient reveals close similarity in the skeletal representation in two assemblages *only* when r is close to 1, the slope is close to 1, and the y-intercept is close to 0. Such a case is shown in Figure 6.5. In this case, the skeletal representation in the two assemblages is nearly identical.

If the assemblages to be compared are not roughly similar in size, the slope will not be close to 1. This means that the ratio of each skeletal element in assemblage A to that element in assemblage B is not constant (Figure 6.6) and the resemblance between the assemblages is unclear. This problem can be dealt with by standardizing the sample size in some arbitrary fashion, but some critics maintain that this procedure will make the value of r closer to 1 than it would have been.

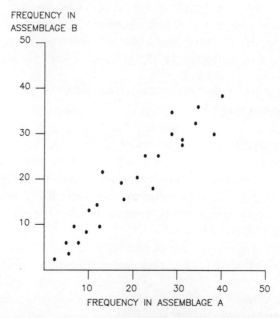

Figure 6.5 Graph of nearly identical representation of skeletal elements in two hypothetical assemblages. Each point on the graph indicates the percentage representation of a single skeletal element in the two assemblages.

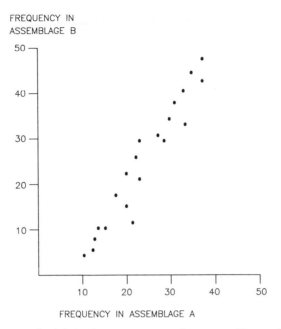

Figure 6.6 Graph of skeletal representation of two assemblages of different size. Although there is a significant correlation between the percentage representation of body parts in assemblages A and B, the frequency of a given part varies in the two assemblages and the meaning of the correlation is therefore difficult to assess.

If the y-intercept is not close to 0, another unclear situation occurs, in which one assemblage contains elements not present in the other. This situation is problematical. Absence of a skeletal element may indicate destruction or loss caused by the prolonged action of the same taphonomic event that has made those elements rare in the other assemblage. In this case, the two assemblages are genuinely similar. On the other hand, the absence of a skeletal element from one assemblage and its presence in the other may reflect quite different taphonomic histories. It is not always possible to distinguish between these alternatives in analyzing assemblages, in which case the researcher is faced with data that are impossible to use. Therefore, even if r is close to 1, if either the slope is not close to 1 or the y-intercept is not close to 0, there is no assurance that the assemblages are, in fact, similar in skeletal representation.

Finally, if the numbers of skeletal elements in each assemblage do not approximate a normal distribution, the value of r may be falsely significant for quite different reasons. Figure 6.7 shows such a case, in which the values of the different skeletal elements fall into two distinct clusters. The value of r is quite high, the slope may be nearly 1, and the y-intercept is close to 0. However, this is clearly a case in which r is an inappropriate measure of the degree of correlation, and the results of this test are meaningless. A spuriously significant value of r occurs more commonly when researchers calculate correlation coefficients on the skeletal representation of an assemblage and that in a whole animal. Because

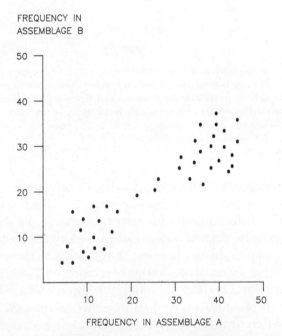

Figure 6.7 Graph of skeletal representation showing an inappropriate use of correlation coefficient. When the sample deviates strongly from a normal distribution, a falsely significant value of r will be derived. As in this graph, a straight line (a significant correlation) can be drawn between any two points or any two clusters of points, but this does not necessarily indicate resemblance.

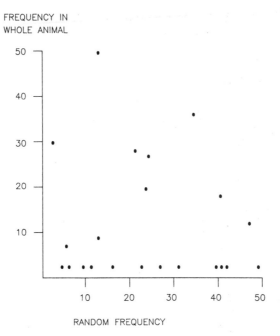

Figure 6.8 Comparison of skeletal representation in an assemblage and in a whole animal. The predominance of paired skeletal elements in a whole animal will cause the value of r to be significant regardless of the frequency of parts in the second assemblage, as is shown by graphing random frequencies of skeletal elements against those in a whole animal. This is a falsely significant correlation coefficient.

many of the skeletal elements in a whole animal occur in pairs, any correlation coefficient will be significant even if the values in the assemblage are random (Figure 6.8). Thus this method cannot be used to determine whether the skeletal representation in a fossil assemblage approximates that in a whole animal.

In summary, correlation coefficients can be used to assess the resemblance between two assemblages, but only in carefully chosen instances. Further, it is probably essential to plot the representation in one assemblage against that in the other, as well as calculating the correlation coefficient, so that spurious results of the kinds shown in Figures 6.6 to 6.8 can be detected. Such graphic representations may highlight the similarities and differences even when the degree of resemblance cannot be assessed statistically.

Relative Abundance of Species

After the specimens have been identified at the first level, the different skeletal elements can be grouped according to taxon to determine the abundance of different species in the assemblage. Because of the differential preservation of animals of different ages and sizes, a straightforward comparison of the numbers of specimens attributed to each taxon is inappropriate. This source of error is especially potent if mammals are compared with nonmammals, because of the differences in the number of skeletal elements in a single individual. For example, although gazelles and crocodiles share a common body plan (four legs, two forelimbs, two hindlimbs, one head, one tail), crocodiles have hundreds of bony scutes in their skin, shed teeth throughout their lives, and have five digits on each hand or foot as compared with the gazelle's two. Thus a single crocodile might contribute many times the number of skeletal elements that could be contributed by a gazelle. Even if only mammals are present, differences in the numbers of digits or tail vertebrae will distort the results. Also, the taphonomic history of an assemblage may have selected against preservation of small species or young individuals even if they were more numerous in life.

For these reasons, the usual procedure is to calculate the minimum number of individuals (*MNI*) represented in an assemblage. The *MNI* is most simply obtained by sorting the skeletal elements of each species into rights and lefts, counting lefts and rights of the same skeletal element as different, and taking the *MNI* as equal to the number of the most common body part. This number can be refined by treating the specimens of adults and nonadults separately and by determining whether or not fragments of the same skeletal element from the same species might be parts of the same original bone. *MNI* is probably the only reasonably accurate indicator of the relative abundance of different species. Of course, a species' abundance in the assemblage may not mimic its abundance in life, depending upon the taphonomic history of the assemblage.

Other useful parameters that may be calculated from the *MNI* are the number of specimens per individual (*NSI*) and the corrected number of specimens per individual (*CSI*). *NSI*, showing the relative completeness of each individual, is derived according to the simple formula proposed by Shotwell (1955:331):

$$NSI = \frac{\text{number of specimens of one species}}{MNI \text{ of that species}} \qquad (6.1)$$

Shotwell (1955, 1958) suggested that relative completeness is probably a function of the proximity of the species' habitat to the site, although this suggestion has been criticized (Voorhies, 1969; Grayson, 1978). The situation is almost certainly more complex than Shotwell initially realized, but his basic insight—that the relative abundance of different species reflects taphonomic history more directly than it reflects their abundance in life—is sound. Thus to some extent species with similar skeletal representations (roughly equal numbers of specimens per individual) may have shared a common or similar taphonomic history. This assumption must be tempered by an awareness that one or more individuals may have arrived at the site as whole or nearly whole carcasses, thus skewing the NSI upward, and by knowledge of the differential vulnerability of different species to destructive forces.

In a later work, Shotwell (1958) pointed out the possible bias inherent in calculating NSI for different species: if the number of elements present in a single individual is substantially different in each species, the NSI's will not be comparable. This problem is minimal when only mammalian species are present, but it becomes a serious source of error if bird, fish, or reptiles are included. Therefore he proposed calculating a corrected number of specimens per individual (CSI) as follows:

$$CSI = \left[\frac{(\text{number of specimens}) \times 100}{\text{estimated number of elements in one individual}}\right] \div MNI \qquad (6.2)$$

Data on relative abundance, as summarized by MNI and CSI, are often presented in pie diagrams. MNI for each species is represented by the length of the wedge on the diagram; CSI is shown by the width of that wedge. As a convention, species are commonly arranged in order of decreasing abundance (MNI), which often correlates with the order of decreasing completeness (CSI). Alternatively, only MNI may be represented in a pie diagram, with the ra-

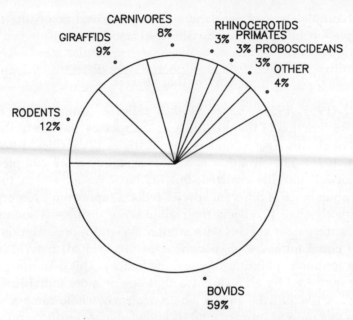

Figure 6.9 Pie diagram showing the relative abundance, as summarized by MNI, of different species in the Fort Ternan assemblage. Each wedge shows the MNI of one or more species. The width of each wedge shows the percentage of the total number of individuals comprised by that taxon.

dial length showing the *MNI* and all wedges being of equal width (Figure 6.9), but it is most efficient to incorporate both sorts of data into a single diagram.

Shotwell (1958) suggested that such diagrams could be used to identify "proximal" and "distal" communities on the basis of their representation in the assemblage. He predicted that the communities that lived nearest the site would show a higher than average *CSI*. Although it is to be expected that animals from distant communities would generally be poorly represented in terms of both *MNI* and *CSI*, studies of modern assemblages show that this generalization and its converse are not always true. For example, many wildebeest drown during river crossings on their annual migration. A resulting assemblage would include large numbers of savannah-dwelling, terrestrial species and few aquatic or riverside species. Yet it would be incorrect to assume that the wildebeest usually live near rivers, even though they are represented by individuals with a high *CSI*. Similarly, in assemblages collected by predators, primates may have lower *CSI*'s than similar-sized antelopes (Brain, 1970). Yet if the assemblage were collected by leopards, which preferentially prey on baboons, primates would be represented by a higher *MNI* than would be predicted from their abundance in the animal community. By examining abundance and completeness data in the light of such knowledge, it may be possible to assess what particular agents of concentration or collection were at work. Further, similarities of representation and abundance among different species may be useful in grouping the species into their original animal communities.

Animal Communities

These sorts of data are often gathered in hopes of delineating the original animal community. A community may be defined as a recognizable, natural grouping of species found in a particular habitat or area. A community often has a distinct trophic (or feeding) structure and a fairly stable pattern of energy flow. In terrestrial communities, energy flows from plants through herbivores to carnivores and then to the microorganisms responsible for decay. In many cases the particular species or types of species in a given type of community are predictable enough that the presence of several species makes it probable that certain others were also present. Finally, it has been shown that modern communities have properties and characteristics *of their own*, in addition to those of each component species. Communities may be defined by the dominant species (an *Acacia tortillis* forest), by the physical habitat (a high-altitude steppe), by the energy pathways (a terrestrial community), or by some combination of these. Thus, identifying the type of community represented in a fossil assemblage may yield much more information than can be obtained by studying each species in isolation.

The importance to paleoecologists of community structure cannot be overestimated. Olson (1961, 1962, 1966) demonstrated that the structure of the food web in an individual community may remain constant over a long period of time, regardless of the extinction or appearance of species and changes in environment. That statement may seem to contradict the common notion that a species is finely tuned and highly responsive to its environment. However, Olson pointed out that the stability of community structure over time means that the pattern of interactions is more important than the presence or absence of any individual species. In fact, Olson proposed that the number of possible community structures was limited during the Permian period—and the same may be true of the Recent. He tied a change of structure, based on a shift in the primary food resource from aquatic to terrestrial plants during the late Cretaceous, to the origin of the mammals. It seems likely that the structure and nature of a food web in a community restricts and defines the niches available to its component species.

Most assemblages contain species from more than one community and habitat, so it is especially important to attempt to sort the fossils into their component communities if the fauna is to be compared with those from other sites. Otherwise, there is a very real possibility that the resemblances between, say, two forest-dwelling faunas will be obscured by the presence of a few savannah-dwelling animals in one of the assemblages. Such inclusions may reflect the agents of concentration at each site more directly than the paleoenvironments.

COMPARISONS OF COMMUNITIES

Broader comparisons between fossil assemblages or between fossil and modern assemblages (or communities) rely on the concepts of species diversity and trophic replacement.

Species diversity. Animal censuses in different modern environments suggest strongly that animal communities may be classified into broad types on the basis of characteristic patterns of diversity (numbers of species and their abundance) and proportions of different trophic types, such as browsers, grazers, frugivores, and primary and secondary carnivores. Work by Bourlière (1963, 1973), Van Couvering and Van Couvering (1976), Andrews, Lord, and Nesbit Evans (1979), and Andrews and Van Couvering (1975) suggests the following characteristics that distinguish forest communities from savannah communities:

1) Tropical forest habitats are typically low in mammalian biomass (weight of living organisms) relative to savannah habitats.
2) Tropical forest animal communities typically show high diversity (many species) of smaller mammals, such as golden moles, shrews, bats, lorises, flying squirrels, and apes. Modern members of these groups are predominantly or exclusively forest-dwelling. Savannah communities, in contrast, show a low diversity of such animals and a high diversity of species of cats, hyenas, dogs, and antelopes (Figure 6.10).
3) The relative abundance of such animals mimics their diversity in species. That is, forest communities are rich in small rodents, primates, insectivores, squirrels, and the like, and each of these types is represented by many individuals; savannah communities are dominated by antelopes and horses and show substantial numbers of medium- and large-sized carnivores. In the absence of these particular taxa, the communities can be expected

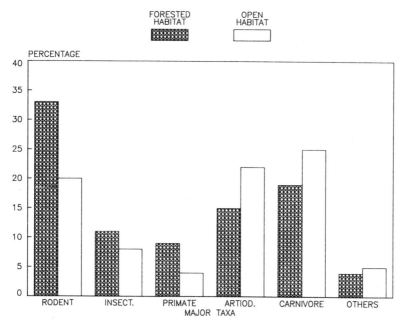

Figure 6.10 Histogram comparing the diversity (percentage of species) of major taxa in an open habitat and in a heavily forested habitat. (Data from Andrews et al., 1979.)

Faunal Analysis / 147

to show comparable proportions of herbivores and predators of comparable size.
4) Forest communities are typically dominated by small animals; savannah communities are typically dominated by large animals.

Although these examples are drawn from Africa, the generalizations hold true elsewhere if similar indigenous taxa are substituted.

In addition, Fleagle (1978) suggested that an animal's size is so fundamental an adaptation that comparing the range of animal sizes in two assemblages may reveal similarities or differences in the niches occupied in each assemblage. Size as an important correlate of many aspects of a species' lifestyle is supported by many types of evidence (Western, 1979). Fleagle confined his comparisons to a single taxonomic family to try to trace changes in niches over time. The problem with this approach is that the niches filled by one family at one time may be encroached upon by species from another family at another time. Such events are masked if comparisons are only intrafamilial; extending this approach to several different families over time may be more fruitful.

By applying these generalizations, researchers may be able to deduce what environment and community is represented in a given assemblage. However, careful attention must be paid to the sedimentary context, the taphonomic history of the assemblage, and the techniques used by researchers to collect the fossils. It is quite possible for the remains of a savannah community to mimic those of a forest community if the taphonomic events removed the remains of all large species and preserved and concentrated only those of smaller animals. Faunal composition per se is only weak evidence of the habitat or community sampled. Further, it is spurious to compare assemblages that have been collected by different techniques. Sieving, in particular, may dramatically alter the data on faunal representation. Assemblages from unsieved sites cannot be expected to resemble closely modern animal communities that include high proportions of small species; sieving almost inevitably raises the proportions of smaller vertebrates represented.

One means of assessing the similarity of two faunas was developed by Simpson (1947, 1960). His faunal resemblance index (*FRI*) is:

$$FRI = 100\frac{C}{N_1}$$

where C = number of taxa in common, and N_1 = number of taxa in the smaller sample. (6.3)

An advantage of this index is that it minimizes the error inherent in comparing a large assemblage with a small one. But as Simpson remarked (1960:305) "It is evident that faunas will be more similar if the taxa in common are relatively abundant . . . and less similar if the most abundant taxa are particular to each." Therefore he proposed a second index, which can be called the relative abundance index (*RAI*), appropriate for comparing two faunas:

$$RAI = \left[\frac{I_c}{I_1 + I_2}\right] \times 100,$$

where I_c = total number of individuals in the taxa in common;
I_1 = total number of individuals in sample 1; and
I_2 = total number of individuals in sample 2. (6.4)

The *FRI* has been used by Van Couvering and Van Couvering (1976) in comparing faunas from different East African Miocene sites with each other and with modern faunas from different habitats. By comparing a series of modern mammal faunas of known habitat, they were able to demonstrate the sensitivity of the *FRI* to habitat differences. Comparisons between two modern forest faunas yielded *FRI*'s of 78–84 for species and 69–83 for genera; these data show the extent to which basically similar habitats differ from identity (*FRI* = 100). In contrast, comparisons of ecologically dissimilar modern areas (forests and savannahs) yielded *FRI*'s ranging from 33–51 for species and 47–62 for genera. The researchers note that time may be a major factor contributing to low *FRI*'s, because there is an increasing chance of habitat change over time and thus of faunal change as well. In addition, there is an increasing change of evolutionary change over time.

Trophic replacement. Another mode of comparison that is more appropriate when the sites are widely removed from one another in time involves the idea of trophic replacement, meaning that a species fulfilling one broad role (such as herbivore, omnivore, or frugivore) in the food web is replaced by another, not necessarily related, species of the same trophic type (Olson, 1961). For example, the extinct Early Miocene species of large herbivores, such as gomphotheres, mammoths, and anthracotheres, have to a large extent been replaced by other large herbivores, such as rhinos, elephants, and hippos. Table 6.2 shows that although the fossil and modern taxa differ widely, the trophic types do not. The number of different trophic types, like the number of species in particular families, can be used to characterize and distinguish forest communities from savannah communities. This concept must be used with caution, however, since there is good evidence (Van Valkenburgh, personal communication) that not all trophic types are present in communities from all time periods.

Extreme caution is necessary in using information about modern community composition to interpret fossil faunas. Certainly the danger of coming to erroneous conclusions is greatest at the oldest sites, where the proportion of taxa that have left no modern descendants is highest. Further, assigning habitat preferences to fossil forms on the basis of modern species' preferences involves a certain risk, lessened perhaps if fossil and modern forms share a suite of habitat-adapted traits. For example, the fact that all modern ostriches live in grassland areas does not preclude the possibility that some ostriches were once woodland or forest dwellers.

In the absence of evidence to the contrary, such assumptions must be made if we are to gain any sense of past faunas and habitats. But it is important to actively seek evidence that habitat preferences or behaviors of extinct species were different from those of living species. Such evidence may be found in the functional anatomy of the extinct form, its associations with other species of plant or animal, or its sedimentary context. As Kay and Cartmill point out (1977:2):

> It might be urged that considerations of physics and mechanics allow us to make many reliable inferences from form without looking for living analogs. This is true for some inferences leading to negative conclu-

Table 6.2 Replacement in trophic types of African Neogene mammals.

Trophic type	Early Miocene group	Degree of replacement	Modern group
Insectivores	Chiropterans Tenrecids[a] Erinaceids Chrysochlorids Soricids	Partial	Chiropterans Erinaceids Chrysochlorids *Soricids*[b]
Anteaters	Tubulidentates	Partial	Tubulidentates Pholidotids
Small Omnivores	Macroscelideans Prosimians	Partial	Macroscelideans Prosimians Cercopithecoids
Large Omnivores	Hominoids Suids	Almost complete	Hominoids (ex *Homo*) Suids Cercopithecoids
Small Herbivores	*Phiomyoids* Anomaluroids * Cricetodontids[c] Sciuroids Ochotonids[a] Procaviids	Almost complete	Phiomyoids Anomaluroids Hystricids *Cricetids* *Murids* Sciuroids Leporids Procaviids
Medium Herbivores	* Pliohyracids *Tragulids* * Gelocids Bovoids	Almost complete	Suids Tragulids *Bovids*
Large Herbivores	* Deinotheriids * Gomphotheriids * Mammutids * Chalicotheriids Rhinocerotids * Anthracotheriids	Almost complete	Elephantids Rhinocerotids Equids Hippopotamids Giraffids *Bovids*
Carnivores	* *Creodonts* * Amphicyonids Viverrids Felids	Almost complete	Viverrids Felids Hyaenids Mustelids Canids

Source: Van Couvering and Van Couvering (1976).
a. Group that no longer lives in Africa.
b. Italics mark taxa that were clearly dominant in trophic groups.
c. Asterisks mark extinct groups.

Some taxonomic groups that vary significantly in body size or feeding habits are entered twice, but the table does not show minor deviations, such as *Proteles* (aardwolf), the anteater hyaenid, or the rabbit-sized bovids. "Large omnivores" and "medium herbivores" are approximately the same size. The term *trophic replacement* does not necessarily mean that the modern groups competitively excluded the ancient groups, although in some instances this may have been the case. It is simply that the fossil record is inadequate to establish the specific cause of the replacement in each case.

sions: there can be no doubt, for instance, that brontosaurs were unable to fly. But positive conclusions based on biomechanical theory need to be tested for living animals before being applied to fossils. Suppose, for instance, that a study of the hand skeleton from Bed I at Olduvai Gorge . . . showed that stresses in this hand would have been maximized in a hanging posture and minimized in a knuckle-walking position. The obvious conclusion, that the animal represented by the fossil hand bones was a knuckle-walker, would have to be rejected if, for example, a similar analysis of a gibbon hand yielded the same conclusion.

It may also be essential to understand the evolutionary context of a given species. As Andrews and Walker (1976:288) observed, it is not reasonable to expect an early member of a particular group to possess all of the characteristics of later members of that group.

INFERRED HABITAT PREFERENCES

Despite the cautions discussed above, data on the relative abundance of different species may provide useful paleoenvironmental information. As Vrba (1980) noted, some basic environmental adaptations of modern groups have a long history. Bovids, in particular, can be useful indicators of paleoenvironments. Since the Middle Miocene or earlier, the caprines (or other aegodonts) and boselaphines (or other boodonts) have shown adaptations to more open and more closed habitats, respectively. Boodonts today are represented by such tribes as the Cephalophini (duikers), Bovini (cattle, buffalo), Tragelaphini (eland, kudu, bushbuck), Boselaphini (nilgai), Hippotragini (roan and sable antelopes, oryx) and Reduncini (waterbuck, reedbuck). Aegodonts are represented by Antilopini (gazelles, springbok), Neotragini (steenbok, oribi), Peleini (grey rhebok), Caprini (goats, sheep), Ovibovini (musk ox, takin), Alcelaphini (wildebeest, hartebeest), and Aepycerotini (impala). Table 6.3 shows the representation of these tribes in modern habitats. Aegodonts show a broad preference for, and an apparently better adaptation to, drier, more open habitats, and boodonts to wetter, more closed habitats. Therefore, the relative abundance of different species of bovids may reveal the types of habitats sampled.

According to Vrba an original community may be represented in an assemblage through either of two sorts of sampling. An X-type sampling results in a relatively undistorted paleoecological

Table 6.3 Percentage representation of extant bovid tribes in some African game parks, reserves, and areas.

Area	Cephalo-pini	Bovini	Tragela-phini	Hippo-tragini	Redun-cini	Neotra-gini	Antilo-pini	Alcela-phini	Aepycero-tini
			Type A: areas with a higher proportion of bush and tree cover						
Kruger	1	11	5	1	2	3		5	72[a]
Manyara		66	1		2				31
Quicama	18	40	28	7	7				
Bicuar	*18*	4	17	*18*	6	13		*18*	6
Luando	14	2	12	39	30	3			
Cuelei	12		17	5	37	23			
Mkuzi	<0.5	15	5		1	<0.5		6	82
Hluhluwe	<0.5	15	21		7			12	33
Kafue	4	13	6	5	*30*	13		24	<0.5
Wankie	5	33	14	7	3	10		28	21
								7	
			Type B: areas with a lower proportion of bush and tree cover						
Kalahari	1		9			2	35	30	
Nairobi	<0.5		3	23	3	<0.5	28	45	21
Lake Turkana				30			24	46	
Serengeti[b]		7	1	1	1		24	58	8
Ngorongoro		<0.5	2				26	71	
Etosha	1		10	19	1	3	48	19	

Source: Vrba (1980). (From *Fossils in the Making*, A. K. Behrensmeyer and A. P. Hill, eds., reprinted by permission of the authors, editors, and the University of Chicago Press © 1980, The University of Chicago.)

a. Maximum percentages per area are in italic.
b. In the case of the Serengeti, migratory, woodlands resident, and plains resident species of Schaller's (1972) table 32 have been added, and his category "Other," containing 10,000 individuals, has been arbitrarily split among roan [antelope] plus *Oryx* (5,000), reedbuck (2,500), and bushbuck (2,500).

picture, because the taphonomic events prior to burial have acted randomly on the components of the original community that might serve as ecological indicators. In such cases, sampling occurs along the gradient of factors that contribute to determining the environment. In a Y-type sampling the taphonomic factors bias an assemblage, so that a group that might serve as an environment indicator is disproportionately included. In other words, the sampling cuts *across* environmental gradients and thus favors one end of the spectrum over the other. The most common case of Y-type sampling is a predator-collected assemblage. Figure 6.11 presents these concepts in graphic form.

Determining which type of sampling has occurred is very important, since an assemblage produced by X-type sampling will be of maximum usefulness in paleoecological reconstructions, and an assemblage produced by Y-type sampling, as by predators, will be of little use for this purpose. Vrba proposed that analyzing the weight distribution of different prey species in the assemblage will

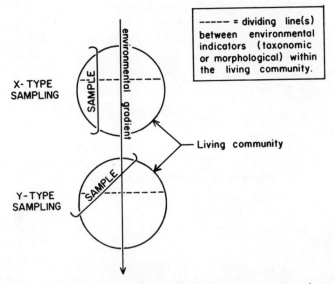

Figure 6.11 Graph of X-type and Y-type sampling of an animal community. X-type sampling provides a relatively undistorted picture of the original community, whereas Y-type sampling is strongly biased. (Courtesy of Elizabeth Vrba.)

show which type of sampling has occurred, based on the observation that a major element in a predator's selection of prey is the size of the prey relative to the predator's size. For predators that have modern analogues or relatives, the preferred size range of prey is fairly easy to document (see Pienaar, 1969). Vrba divided bovids into four weight classes: up to 50 pounds, 50–200 pounds, 200–650 pounds, and above 650 pounds. An assemblage that includes prey in only one weight class is likely to be too heavily biased by the predator's preferences to be useful in paleoecological reconstruction (Y-type sampling). If the sedimentary context indicates that hydrodynamic sorting influenced the faunal representation, a restricted range of sizes of prey species would probably also reflect Y-type sampling. An assemblage that samples even two adjacent weight classes may be a more reliable indicator of the ancient environment.

Vrba (1975, 1980) also suggested a model for distinguishing between assemblages accumulated by primary predators (primary assemblages) and those accumulated by scavengers and other secondary collectors such as porcupines (secondary assemblages). To use this model, the necessary data are: the relative abundance of all bovid (or other prey) species in the assemblage; the estimated live weights of those species (which need be accurate only to the broad weight classes given above); and the numbers of individual adults and nonadults (here designated as juveniles) for each species. Using these data, the characteristics of an assemblage can be compared to those of primary and secondary assemblages (Table 6.4). To assess what a high percentage of juveniles might be, Vrba cited data on the proportion of juveniles in various living bovid species, which ranges from about 20 percent to 50 percent. Given that juvenile skeletons are more vulnerable to destruction from nearly all taphonomic events, any fossil assemblage with a percentage of juveniles comparable to that in living populations is likely to be a primary assemblage.

One kind of secondary assemblage that can be recognized from the fossil record is the porcupine-collected assemblage. Recognizable largely on the basis of skeletal representation and gnawing marks, such an assemblage may provide a very accurate sampling of the local animal community. Brain (1980) compared the *MNI* of

Table 6.4 Model for interpreting history of bone accumulations using adult-juvenile and weight data.

Stage 1
Bovids are killed by predation in the vicinity of the cave or die a natural death in or near the cave.

Stage 2
Remains of dead animals are brought into the cave to eventually constitute:

Primary assemblages	*Secondary assemblages*
Those brought into the cave, which serves as a lair or shelter, by the primary predators (hominid hunters are included as a subset of the set of predators throughout).	Those brought into the cave by scavengers or collectors.
Unlikely to contain a low percentage of juveniles.	Likely to contain a low percentage of juveniles.
Most individuals fall into a restricted body weight range (only where one predator or predators preying on prey of similar size, predominate(s)).	The body weight distribution may have a high variance.

Source: Vrba (1980). (From *Fossils in the Making*, A. K. Behrensmeyer and A. P. Hill, eds., reprinted by permission of the authors, editors, and the University of Chicago Press © 1980, The University of Chicago.)

different species represented in the Nossob porcupine assemblage with the relative abundance of different species known to live in the area. If the only species considered are those for which both census data from the live population and data on representation in the fossil assemblage are available, the Nossob sample is highly correlated with the animal community ($r = .9329$ with 4 degrees of freedom; $p < 0.01$); see Figure 6.12. This means that a fossil assemblage known to have been collected by porcupines can be used with confidence in reconstructing the paleoenvironment and the ancient animal community.

Population Dynamics

Once the relative abundance of different species has been established, the population dynamics of the most abundant species can be analyzed. Information on the size and age structure of an ani-

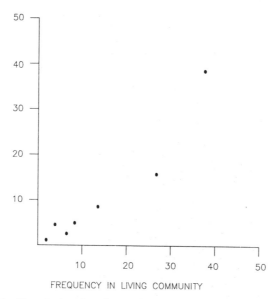

Figure 6.12 The relative abundance of species in the Nossob porcupine lair plotted against the relative abundance of species in the living animal community. (Data from Brain, 1980.)

mal population can be especially useful in tracing the taphonomic history of a fossil assemblage. Studies of population structure may yield clues about the basic mortality pattern, season of death, and time interval between death and burial. Klein (1978) has also shown the usefulness of such work in establishing hunting patterns among ancient hominid populations. There is little to be gained from such analyses unless the *MNI* for the species is sufficiently large that members of different age classes are likely to be represented. Species represented by fewer than thirty individuals are probably not suitable for such analysis.

Most of the work done on the population dynamics of fossil populations (such as Van Valen, 1964 a, b; Voorhies, 1969; Clark and Guensburg, 1970; Saunders, 1977) borrows heavily from classic studies of population dynamics in living animals, like that of Deevey (1947). Using what is now a standard life table, Deevey recorded the age structure and mortality rates of a population of Dall mountain sheep (*Ovis dalli dalli*), based on data from a long-term behavioral study by Murie (1944). Such life tables summarize the vital statistics of mortality, survivorship, and life expectancy for each class in population. Differences in absolute size of populations are eliminated by standardizing the life table to a hypothetical population of 1,000 individuals.

Kurtén (1953) explored the applications of such analytical techniques to fossil assemblages. He pointed out the effects of different mortality patterns (attritional versus catastrophic) on the sample of a population represented in a fossil assemblage and proposed a different mode of analysis for each pattern. If preserved, the remains from a catastrophic mass death samples all members of all populations within the area where death occurred, because the cause of death is unrelated to the age of the animals. For such cases, Kurtén proposed constructing a time-specific life table. The raw data (numbers of individuals in each age class) are assumed to represent the survivors at the beginning of each age interval (l_x). The number dying during each interval (d_x) is calculated by subtracting the number in each age class from the number in the next youngest age class; that is, d_x is the shrinkage between age class a and age class $a +$ one. Age classes, rather than years of age, are used because it is often difficult to establish the age of an individual precisely.

The second mortality pattern, attrition, produces a true sampling of the death rates in different age groups over time. If preserved, populations showing attritional mortality will have high death rates in the most vulnerable age classes (usually very young and very old) and low rates in the least vulnerable classes. Population biologists, who can follow a cohort over time, use a dynamic or horizontal life table. Kurtén suggests a modification of this, the composite life table, for use with fossil assemblages that probably reflect attritional mortality rates in successive populations over a long period of time. In a composite life table, the number of individuals in each age class is assumed to be the mortalities (d_x) for each age class rather than the survivors at the beginning of each age interval (l_x). To calculate l_x in such instances, the d_x is subtracted from l_x of the preceding group, starting with the total number of individuals, standardized to 1,000 (Table 6.5).

Table 6.5 Composite life table for a hypothetical species.

Age class	l_x[a]	d_x	$1,000q_x$	e_x
1	1,000	160	160	3.31
2	840	140	166	2.85
3	600	50	83	2.78
4	550	60	109	1.99
5	490	200	408	1.17
6	290	250	862	2.97
7	40	40	1,000	.5
8	0	—	—	—

a. l_x indicates the number of individuals that are alive at the beginning of the time interval represented by one age class; d_x is the number of individuals that die during that interval; $1,000q_x$ is the mortality rate per thousand; e_x is the mean life expectancy of the individuals that are alive at the end of the interval. Age classes, representing time intervals of indeterminate length, are used because the age of some species can be determined very precisely, as when there is a biannual increment of dentine, and the age of others can be determined no more accurately than a two- or three-year time span.

Plotting survivorship curves is another common mode of analyzing population data that is applicable to fossil populations. The data are based on the life table for a given population. Survivorship curves result from plotting the survivors (l_x) against age classes or time intervals, when the age classes can be translated into time intervals. Figure 6.13 shows three basic types of survivorship curve among mammals. Curve A is convex, indicating that most of the mortality in the population occurs at the end of the lifespan. Such a pattern may occur in K-selected species that raise few young but that make heavy parental investment in each offspring (MacArthur and Wilson, 1967). Curve B, the diagonal, shows a mortality rate that does not vary as a function of age. Kurtén (1953) noted that a diagonal survivorship curve would be expected from catastrophic mortality. Curve C, the concave type, shows a population in which mortality is very high among nonadults, relatively constant and low among adults. Species showing this curve are classic r-selected species that produce large numbers of young to ensure the survival of adequate numbers to reproductive age (MacArthur and Wilson, 1967). Thus the type of survivorship curve may reveal the basic survival strategy of a species and the mortality pattern that characterized its death.

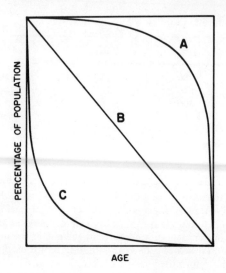

Figure 6.13 Three basic types of survivorship curves for mammals. Type A is convex, indicating high mortality among aged individuals; type B is diagonal, indicating that mortality is independent of age; type C is concave, indicating high juvenile mortality. (Reprinted from *Fundamentals of Ecology*, third edition, by Eugene P. Odum. Copyright 1971 by W. B. Saunders Company. Copyright 1953 and 1959 by W. B. Saunders Company. Reprinted by permission of Holt, Rinehart and Winston.)

There are several basic problems in analyzing the population dynamics of fossil assemblages. First it is necessary to establish that a population has been sampled. A true population is represented in the fossil record only in cases of mass catastrophic death. Such occurrences are relatively rare, yet they are highly likely to be preserved if the mode of death is conducive to fossilization, as in deaths caused by flood or by volcanic ashfall. Attritional mortality, as preserved in the fossil record, will show average mortality rates for successive populations over time. But as Raup and Stanley (1971) conceded, if mortality rates and population growth rates are more or less stable over time, it is unimportant whether the presumed population is an interbreeding, contemporaneous population or a sequence of populations.

The second major problem in analyzing fossil population data is that even in abundant species, the fossil record may not accurately reflect the living population. Fossil assemblages are generally biased against nonadult animals, because the bones of juveniles are more vulnerable to postmortem destruction. Although some attritional assemblages may accurately reflect the proportions of different species in the original community, the population structure of each species is distorted. Catastrophic mortality provides a reliable sampling of all age classes only when it is followed by immediate burial, with little or no time for the destruction of bones by weathering, breakage, or chewing by carnivores.

The third difficulty in analyzing fossil populations is determining the ages of the animals in the sample. The literature on various techniques of age determination is extensive and is reviewed by Klein (1978), Spinage (1973), and Phillips-Conroy (1978). In general, age can be determined most objectively in species in which annual or seasonal increments of bone, dentine, shell, or cartilage can be discerned. The most subjective age determinations rely on the criteria of growth rates or wear of various structures; unfortunately, both growth and wear rates are notoriously variable in different climates, environments, and seasons, in populations with different densities, and with other kinds of variation (Stanley-Price, personal communication).

Klein (1978) proposed a modification of the standard techniques of assessing rates of wear on teeth, based on measurements of crown height of isolated bovid molars and premolars (Figure 6.14). The crown height data were converted to age classes by: (1) determining the average height of an unworn crown for each tooth; (2) determining the maximum lifespan for each species by searching the literature on modern bovids; (3) establishing a rate of wear by dividing the unworn height by the maximum age of an animal of that species; and (4) grouping data into age classes on the basis of years of wear that had occurred. Obviously, this technique is most suitable for still-living species and those with well-studied modern relatives. Because bovids are often abundant in the fossil record, this technique may be of great value in establishing age classes for such species. Problems may occur in comparing age distributions from widely separated sites, because tooth wear varies with environment. Fortunately, among young individuals, age classes can be defined by the eruption of various teeth; among

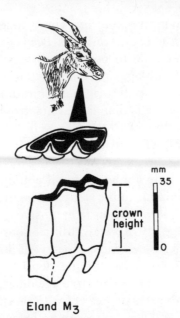

Figure 6.14 Crown height as measured on an isolated eland molar. (From Kelin, 1978; reprinted by permission of the author and publisher. Copyright by Academic Press, Inc. [London], Ltd.)

adults, the ease or difficulty of age determination depends on the mortality pattern.

The final difficulty in analyzing the population dynamics of fossil species is assessing the effect of the mortality pattern on the representation of the original population. The first step is to determine whether the assemblage resulted from catastrophic or from attritional mortality. Voorhies (1969) suggests comparing the age distribution of the population in question with the age distributions characteristic of catastrophic and attritional mortality, as summarized in Figure 6.15. The figure shows a histogram of the age distribution of a hypothetical assemblage derived from a living ungulate species population of stable size; in the assemblage, nonadults would make up a greater proportion of the population than adults because mortality is highest among young animals. Very old animals also have a high mortality rate, while adults in their prime have a relatively low mortality rate. The age structure

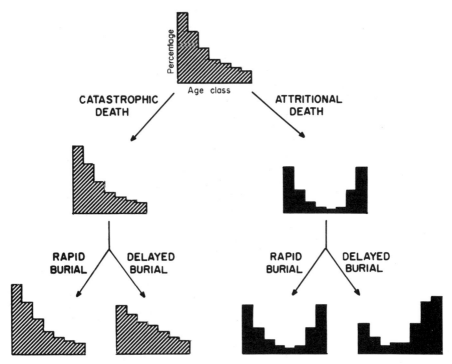

Figure 6.15 Changes in the age structure of a hypothetical ungulate population according to mortality type and taphonomic history. (Drawing by Dave Bichell.)

of an assemblage from this original population would depend on the mortality pattern and on whether burial was immediate or delayed. The major difference between the structure of fossil populations resulting from catastrophic and attritional mortality is the peak of deaths among adults in the latter case. In either case, large numbers of nonadults die, although they may be removed from the fossil record by taphonomic processes if there is sufficient time lag between death and burial. Once the mortality pattern has been identified, the age class data can be used in the appropriate life table and transformed into a survivorship curve.

Two examples of fossil population dynamics will demonstrate

the usefulness of and problems inherent in these techniques. Voorhies (1969) studied the remains of an antilocaprid, *Merycodus furcatus*, from the Verdigre Quarry in Nebraska. The 475 *Merycodus* individuals represent 83 percent of all individuals in the assemblage. The age structure of this species was determined by grouping the 1,100 mandibles into seven groups on the basis of tooth eruption and wear (Figure 6.16). The groups are remarkably distinct, with few or no intermediates between age classes. In addition, the crown heights of two molars show a multimodal distribution that corresponds with the age classes (Figure 6.17). Voorhies interpreted these sharply defined age classes as evidence of seasonal births. He also suggested that this pattern would not be seen if mortality were attritional; only catastrophic death or deaths occurring at the same season for several years would preserve the distinctness of the age classes. Further, the age distribution closely

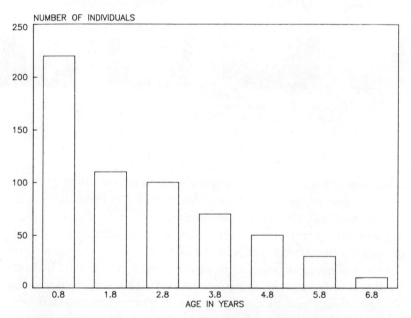

Figure 6.16 Age structure of the population of *Merycodus furcatus*, an antilocaprid, preserved at the Verdigre Quarry, Nebraska. (From Voorhies, 1969; reprinted by permission of the author and *Contributions to Geology*.)

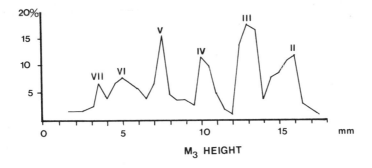
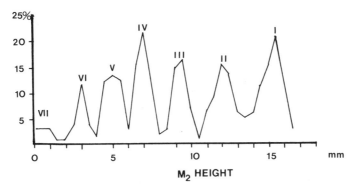

Figure 6.17 Frequencies of heights of the metaconid cusp on M_2 and M_3 of *Merycodus furcatus* in the Verdigre assemblage. Roman numerals denote age classes, which are sharply defined in this population. (From Voorhies, 1969; reprinted by permission of the author and *Contributions to Geology*.)

resembles that seen in present-day ungulate populations (compare Figures 6.15 and 6.16). Only catastrophic mortality samples individuals from all age classes in the proportions in which they would be expected to occur in life.

In this case differential destruction of the remains of individuals in different age classes may have occurred. But even if it is assumed that the youngest age classes are badly underrepresented because of differential destruction, the age distribution still resembles that expected from catastrophic mortality. But as in all such studies, Voorhies' interpretation could be correct *only* if the original population were stable and had a high proportion of young. A waning population in a disrupted habitat might show a very different age structure.

Klein (1978) studied special types of mortality patterns among the bovids from the Stone Age sites at Klasies River Mouth and Nelson Bay, South Africa. Both sites are believed to preserve the faunal and archaeological remains of hominid hunting; their proximity and their similarities in age and artifacts led Klein to treat the two sites together. Both faunas are rich in bovids, whose population dynamics were analyzed in hopes of finding out the hominids' hunting strategies and prey selection criteria. Bovids were grouped into age classes on the basis of crown heights, as discussed above. Klein found two statistically different patterns of age distribution in two species groups.

To determine that the age distributions were different, Klein used the Kolmogorov-Smirnov test (Goodman, 1954). This test is sensitive to the ranking of classes (in this case, youngest to oldest) as well as to the abundance within each class. The formula for the Kolmogorov-Smirnov test is:

$$\frac{\text{Maximum PD}}{\sqrt{\frac{(n_1 + n_2)}{n_1 n_2}}} \quad (6.5)$$

where maximum PD = maximum percentage difference; n_1 = sample size assemblage 1; and n_2 = sample size assemblage 2.

Maximum percentage difference is derived by expressing the abundance in each age class in sample 1 and 2 as cumulative percentages. The cumulative percentage in each age class is then subtracted from the cumulative percentage for that age class in the other assemblage. The maximum difference for any age class then becomes the maximum percentage difference. In this case, the maximum percentage difference is 46, which gives a value of 3.07 for the Kolmogorov-Smirnov test. Any value of 1.63 or above is considered significant at the level of $p < 0.01$ for this test; thus the age distributions of these two bovid groups are significantly different.

In the first group (Cape buffalo, giant buffalo, blue antelope, and roan antelope), most of the animals killed were in the first 10 percent of their lifespan. In contrast, young individuals in the second group were killed relatively rarely; their age distribution shows a predominance of prime adults. Both patterns of age distribution are found at both sites, so the differences must be attributed to differences in the hunting strategies employed. Klein presented the age distribution data in what he called "predation curves," which are similar to survivorship curves except that they represent the animals that die rather than the survivors. The slope of the line for any age class shows the intensity of predation on that class: the steeper the slope, the more individuals died of predation by hominids (Figures 6.18 and 6.19).

Klein interpreted the differences in age distribution of the two groups as reflecting different hunting strategies. The age distribution of the first group, which is superficially similar to that from

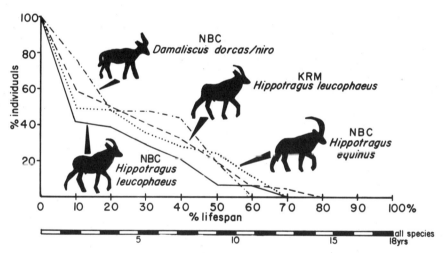

Figure 6.18 Predation curves for the bastard hartebeest (*D. dorcas* or *D. niro*), the blue antelope (*H. leucophaeus*), and the roan antelope (*H. equinus*) from Nelson Bay Cave (NBC) as compared with that for blue antelope from the Middle Stone Age levels at Klasies River Mouth Caves (KRM). The bastard hartebeest was hunted most heavily later in life than the other species. (From Klein, 1978; reprinted by permission of the author and publisher. Copyright by Academic Press, Inc. [London], Ltd.)

Faunal Analysis / 167

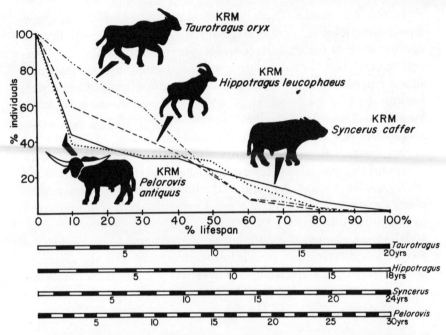

Figure 6.19 Predation curves for the eland (*T. oryx*), the blue antelope (*H. leucophaeus*), the Cape buffalo (*S. caffer*), and an extinct species of giant buffalo (*P. antiquus*) from the Klasies River Mouth Caves. The eland was hunted more heavily later in life than the other species. (From Klein, 1978; reprinted by permission of the author and publisher. Copyright by Academic Press, Inc. [London], Ltd.)

catastrophic mortality, suggests that the hunters were selecting the most vulnerable individuals: the young. Such a pattern is seen among nonhuman predators when they hunt singly. The second group, in which large numbers of prime adults were killed, reflects a different sort of hunting strategy. Klein suggested that these species were hunted in some way that did not depend on the ease of catching or killing particular individuals. He suggested that groups of hunters might have driven the prey off cliffs or into bogs, for example. It is relevant that today both eland and hartebeest live in two sorts of social groups, one consisting of a single male with females and young, the other consisting of bachelor males. Klein's interpretation might agree well with these facts if the hunters more often found male bachelor herds than herds of young and females.

The most interesting point about Klein's interpretations is that the first group of species shows an age distribution pattern that suggests catastrophic mortality but almost certainly represents a specialized form of attritional mortality, in which young were the favored prey but were not killed all at one time. The second group shows the converse; its age distribution mimics that of attritional mortality, in that many adults are killed, but it more probably represents a specialized type of catastrophic mortality. A mass death occurred but the population killed had a very limited age distribution. These examples should indicate the care that must be taken in using age distribution data to determine mortality pattern. All available information on the type of site and species' behavior must be integrated if the interpretation is to be meaningful.

Summary

Faunal analysis can yield several sorts of useful information. Identification of skeletal element and species provides the essential data for further analysis, and important implications may also be drawn from indeterminate remains.

The relative abundance of different skeletal elements can identify the agent that was probably responsible for the formation of the assemblage. Evidence for hydrodynamic sorting (by Voorhies Group) or for selection of particular skeletal elements by some living animal can be gathered by synthesizing data on skeletal representation and the state of preservation of the remains.

The relative abundance of different species within the assemblage may help reveal the characteristics of the original animal community as well as its environment. Such endeavors are most likely to be successful if the taphonomic history of the assemblage can be reconstructed with confidence. The habitat preference of extinct species can be inferred in two ways: by spatial association with a taxon whose habitat preference is known, and by indirect evidence of the habitat as judged from the physical adaptations of the species, in terms of limb or tooth structure or some other characteristic. Occasionally, a sedimentary context will imply a habitat, as when lacustrine or desert sediments are found. However, such information is useful only if the assemblage is known to have been locally derived.

Analysis of the population dynamics of the more abundant species may reveal the basic mortality pattern, season of death, and time interval between death and burial, and may contribute to an understanding of the mode of death itself.

7 Postdeposition and Postfossilization Distortion

The previous chapters have described the various processes that occur after the death of the animal and before its burial and fossilization. The distortions resulting from such postmortem events can, in many cases, be recognized and perhaps corrected for, using the methods and principles I have described. However, the potential for distortion through preferential destruction—or preservation—does not stop once a bone or tooth has been buried in sediments.

Diagenetic Factors

After deposition and fossilization have occurred, an assemblage or an individual fossil may be distorted by changes in the sedimentary rocks that enclose it. Such changes in the rocks and fossils are called diagenetic; this term usually excludes crustal movements and weathering of the rock. The mineralogy and fabric of the rock may undergo extensive change to a point approaching metamorphism, at which stage differences in the composition and structure of the rock are obliterated.

Fossils or bones in sediments may undergo several types of change during diagenesis. Minerals may be precipitated in the voids of the skeletal remains left by the decay of organic materials.

This process, called impregnation or permineralization, may result in the preservation of exceptionally fine details in fossils. The microscopic structure of bones or teeth may be clearly preserved, despite the obvious fragility of such small structures. Horizons containing many skeletal elements may be more permeable, which leads to unusually rapid permineralization (Rolfe and Brett, 1969). The original mineral matrix of bones and teeth may also be replaced by other minerals. It now seems unlikely that such replacement occurs in the molecule-by-molecule method that was once widely accepted (Rolfe and Brett, 1969), although as in impregnation, fine structures may be preserved.

Deformation of various types may occur during replacement, altering the volume, shape, or linear dimensions of a fossil without obvious signs of cracking or breaking. Compaction commonly occurs as sediments are deposited; the weight of the overlying sediments compresses the sedimentary particles (including bones) and lessens the spaces between the particles. When deposition is rapid, this pressure often crushes or breaks bones or fossils; slower deposition may produce plastic deformation, in which the shape or dimensions of the original matrix are changed without breaking. Finally, faulting or erosion may expose sedimentary rocks and their enclosed fossils to weathering. Diagenetic effects on fossils can be grouped into the two broad categories of (1) breakage and crushing and (2) plastic or other deformation.

BREAKAGE

Postfossilization breakage has been recognized at many sites by various workers. It is particularly important to distinguish among breakage that occurred during the lifetime of the animal, breakage after its death, and breakage after its bones were fossilized. On occasion, breakage that was almost certainly postmortem has been taken as evidence of the cause of the animal's death. Such confusion has led to unusually vivid and gory reconstructions of the behavior of our ancestors (for example, Dart, 1957; Ardrey, 1961, 1976), despite the existence of alternative explanations (see Roper, 1969; Walker, 1976a).

Bone breakage that occurred during the individual's lifetime is readily recognized by the signs of healing on the edges of the breaks; growth of new bone leads to rounding and smoothing of the angular surfaces. Such healing may be detectable as soon as a week or two after the injury, which means that in a living animal all but immediately fatal injuries to bone will show healing. Break contours on living or fresh bone either follow the long axis of the collagen fibers or, if the breaks are not quite parallel to the fiber direction, the contours are irregular and splintered.

In contrast, dried bones show a greater tendency to shear perpendicular to the long axis of the bone and its collagen fibers, because of the decreased shearing strength and energy-absorbing capacity of dried bone in that dimension (Evans, 1973). Dried bones that have been trampled may show what have been called columnar fractures (Gifford, 1978), as shown in Figure 7.1. Such fractures result in many rectangular or almost rectangular fragments of bone.

Bonnichsen (1978) reported differences in the breakage patterns of mineralized and fresh bones when struck with stone tools:

Figure 7.1 Columnar fracture of a long bone that has been trampled by a cow; the cow's hoofprint is outlined in white.

1) Fresh bone shows negative impact scars on the lateral edge of the fragment directly below the impact point; these are not present in mineralized bones.
2) Fracturing of mineralized bones produces rectangular fragments (Gifford's "columnar fractures"). These are apparently produced by propagation of dessication cracks parallel to the long axis of the bone as observed at 2000× magnification.
3) Fresh or green bones commonly show spiral fractures, which do not occur in mineralized bone.
4) Fracture surfaces of mineralized bones are judged to be rougher than those of fresh bones.

Many but not all of the characteristic patterns seen in mineralized bones are also seen in dried bone. Dessication cracks and columnar fractures are commonly observed in dried bone, as are spiral fractures, which do not occur in mineralized bone. Although spiral fractures have been taken as evidence of hominid activities, increasing evidence shows that they are also caused by carnivores (Shipman and Phillips-Conroy, 1977), natural weathering (Shipman, 1979), and trampling (Myers, Voorhies, and Corner, 1980).

The fracture surfaces of a spiral fracture produced by weathering and one produced by experimental manipulation of dried bone were inspected using the scanning electron microscope (Shipman, 1979 and in press); see Figure 7.2. Natural weathering or breakage of a fresh tibia produced a spiral fracture that follows the spiral course of the collagen fibers, which can be seen as longitudinal stringlike structures on the fracture surface. An artificially produced spiral fracture against the grain of the fibers showed a distinctive pattern of abrupt microscopic steps on the fracture surface. These differences in fracture surface morphology mean that it is possible to distinguish two types of spiral fractures. A type I fracture follows the predominant direction of the collagen fibers and may have been caused by weathering, trampling, hominids, or carnivores. A type II fracture cuts across the fiber direction and must have been caused by hominids or carnivores capable of exerting sufficient torsion. Further research is needed to determine whether the proportions of type I and type II fractures vary in hominid-broken and carnivore-broken assemblages.

Mineralized bones do exhibit one type of break that is not observed in fresh or dried bones: the perpendicular smooth fracture,

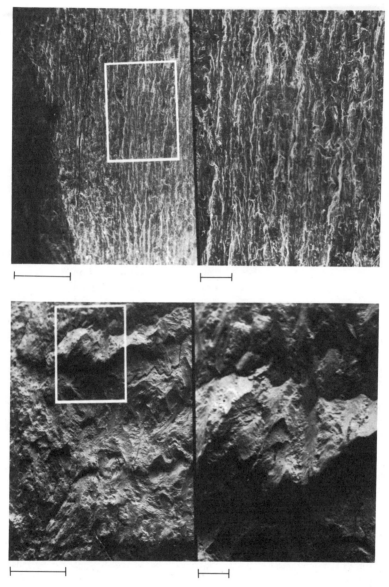

Figure 7.2 Top, type I spiral fracture on a topi tibia, produced by weathering. The laminae of the bone can be seen as parallel structures separated by canals for blood vessels. At right is a close-up of the area in the box. Each bar represents 1 mm. *Bottom,* type II spiral fracture on a topi tibia, produced by torsional stress on the bone at right angles to the predominant direction of the collagen fibers. The stepped and roughened appearance of the fracture surface is caused by the intersection of the fracture front with the laminae. At right is a close-up of the area in the box. The bars represent 1 mm each.

in which the fracture surface is very smooth and flat and is perpendicular to the long axis of the shaft and to the collagen fibers (Figure 7.3). Such breaks probably occur when the fossilized bones respond to diagenetic pressures as if they were part of the rock; the plane of breakage probably corresponds to the cleavage plane of the rock. Further evidence that the break occurred after fossilization can be gleaned from the lighter color of the apparently less

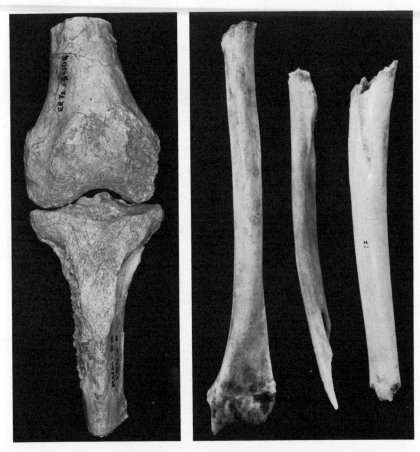

Figure 7.3 *Left*, postfossilization fracture on the distal femur and proximal tibia of a cercopithecid from Koobi Fora, Kenya. Note the smooth, flat surface of the fracture which is nearly perpendicular to the long axis of the bone. *Right*, prefossilization fractures on fresh bones. Note the jagged edges of the fracture surfaces and their oblique angle to the long axis.

mineralized surface of the break (Bonnichsen, 1978) compared to the color of the external bone surface. This difference is often quite marked and can be used to identify postfossilization breaks. Similarly, postfossilization scratches or other marks expose the light interior bone color in most cases and do not contain even microscopic amounts of matrix, as do older scratches.

The only type of prefossilization break that might be confused with a postfossilization perpendicular smooth break is one that results from burning. Bones exposed directly to flame during cooking or other burning processes show a distinctive pattern of breakage (Shipman, 1979 and in press; Stewart, 1979). It is well known that collagen fibers denature at high temperatures. Because collagen is normally under tension in the bone, this denaturing causes transverse cracks to develop perpendicular to the direction of the fibers and the long axis of the bone. In several minutes of burning, some cracks may deepen sufficiently to produce a break across the shaft. The transverse break may transect the entire shaft, or it may intersect with a longitudinal crack and break off a quarter of the bone (Figure 7.4). Although this type of break may resemble a perpendicular smooth (postfossilization) break, the condition of the articular surfaces should distinguish between the two. Burned bones show a characteristic polygonal cracking pattern on subchondral articular surfaces; sometimes the polygonal plates curl up at the edges and eventually fall off. This pattern is distinct from the mosaic cracking of articular surfaces produced by weathering (Figure 7.5). In addition, bones exposed to burning show several

Figure 7.4 The propagation of both longitudinal and transverse cracks has quartered this sheep phalanx after fifteen minutes of exposure to direct flame.

Figure 7.5 Top, scanning electron micrograph of the articular surface of a burned sheep phalanx. Note the concave polygonal plates. *Bottom*, scanning electron micrograph of the articular surface of a carnivore humerus, showing mosaic cracking from weathering. The polygonal structures are separated by broad valleys. Each bar represents 1 mm.

microscopic transverse cracks on the shaft, which will not be present if the break has occurred after fossilization. When examined under a microscope at high power, the apparently smooth surface of the transverse break in burned bones shows some melting of the crystallites.

Postfossilization crushing, as from trampling, rockfalls, or the pressure of overlying sediments, may be difficult to distinguish from crushing of dried bones before fossilization. Again, however, bone surfaces exposed after fossilization should be lighter in color than the rest of the bone.

PLASTIC DEFORMATION

Plastic deformation is widely known from many invertebrate and vertebrate fossils. It is most easily recognized by deviations in the shape of well-known species or by deviations from symmetry. Most of the work on deformation of fossils has been done on invertebrates, which are commonly used to detect deformation in sediments (Ramsay, 1967).

One common type of deformation is called *homogeneous strain,* in which all lines or areas of bone in one particular orientation are uniformly lengthened or shortened. The volume of the bone is unchanged, parallel lines remain parallel, and straight lines remain straight. The effect is one of drawing the fossil on a square grid and then stretching the grid in one direction until it is a rectangle. In such cases, the original shape of the fossil can be approximated if the strain ellipse can be determined. The strain ellipse is the shape that would be produced in a circle subjected to the same homogeneous strain. Once the strain ellipse is known, the magnitude and orientation of the maximum and minimum axes of elongation can be calculated and used to estimate the original shape of the fossil. Ramsay (1967) discussed this procedure more fully.

Not all deformation is the result of homogeneous strain on objects that are essentially two-dimensional. One famous example of plastic deformation combined with crushing is the innominate of the famous "Lucy" skeleton of *Australopithecus afarensis.* Since both a sacrum and a nearly complete innominate are preserved from this specimen, mirror-imaging was used to create a complete pelvis (Figure 7.6). When this was initially attempted, the two pubic rami failed to meet at the pubic symphysis—surely not the

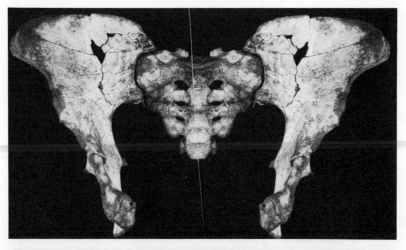

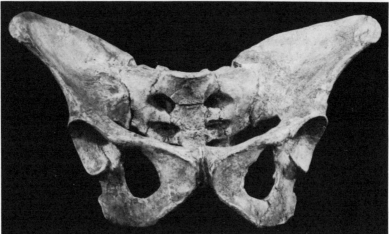

Figure 7.6 *Top*, mirror-imaged pelvis of an australopithecine nicknamed Lucy. Distortion of the pelvis is shown by the fact that the innominates fail to meet at the pubic symphysis. *Bottom*, Lucy's pelvis after reconstruction. (Photos courtesy of C. Owen Lovejoy.)

situation in life. To correct the problem, a detailed, hollow plastic cast was made of both surfaces of the innominate. Each tiny piece was cut out, following the cracks produced by crushing, and the original alignment of the pieces on each surface was recreated as nearly as possible. When the two surfaces were glued back together, some distortion due to plastic deformation remained; by adjusting some of the joints between pieces, a closer approximation of the original shape was achieved. This restoration was exacting but relatively straightforward, because most of the distortion was produced by crushing. In other cases, restoration of the original shape of a fossil is more difficult. Several skulls, such as the *Proconsul africanus* skull from Rusinga and the *Paracolobus chemeroni* skull from Baringo, Kenya, lack bilateral symmetry. Only complex mathematical modeling of the axes of deformation in three dimensions can reveal the original shape of such skulls.

A different sort of deformation occurs when elements of a skull rotate in different directions, as in the KNM-ER 1470 skull from East Turkana, Kenya. In this specimen the position of porion—the uppermost lateral point in the margin of the external auditory meatus—relative to skull length is different on the right and left sides. The difference is substantial enough to place the specimens within the range of *Homo* species on the right and *Australopithecus* on the left (Walker, 1976b; Wood, 1976). Careful plotting in three dimensions of analogous points on the right and left sides showed that the right temporal has been rotated downward and forward from its original position (Walker, 1980); see Figure 7.7.

Another form of postfossilization distortion occurs when matrix infiltrates fine cracks in the original bone, expands, and ultimately splays pieces of the specimen apart. Although the matrix filling in such cracks may be only a few millimeters in width, it may seriously alter the shape of the bone. Attempting to remove such matrix is often technically difficult or unjustifiably dangerous to the specimen. Restoration can often be successfully accomplished by making hollow plastic casts of the fossil in which the matrix can be carefully delineated and then cut away. The remaining pieces of the cast, representing the original bone fragments, can then be glued together to show the original shape. This approach has been used with the palate of KNM-ER 1805, in which the teeth were splayed far out of their original position by matrix (Figure 7.8).

Erosion

Another source of postfossilization bias occurs when a fossil is eroded out of its matrix. Unless the fossil is collected shortly after exposure by erosion, it may fragment or become reworked in a sec-

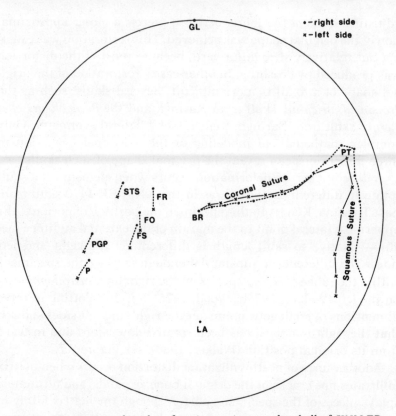

Figure 7.7 Equiangular plot of various points on the skull of KNM-ER 1470. The coronal suture shows symmetry, but the squamous suture on the right side has shifted anteriorly. Various anatomical landmarks are shown: GL = glabella; BR = bregma; LA = lambda; PT = pterion; P = porion; PGP = postglenoid process; FS = foramen spinosa; FO = foramen orale; FR = foramen rotundum; STS = sphenotemporal suture. (From Walker, 1980; reprinted from *Fossils in the Making* by A. K. Behrensmeyer and A. P. Hill, eds., by permission of the author, the editors, and the University of Chicago Press. © 1980 the University of Chicago.)

ond depositional environment. The rate at which a fossil breaks up after exposure is a function of the degree of replacement or permineralization that has occurred and of the natural forces to which the fossil is exposed.

One particularly unfortunate example of the results of erosion and exposure is the fate of Olduvai Hominid 12, known as George. This specimen apparently lay on the surface of the VEK site at Ol-

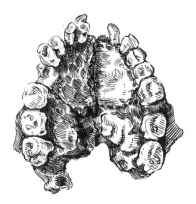 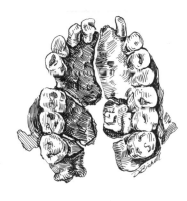

Figure 7.8 The palate of a hominid, KNM-ER 1805, before reconstruction (*left*) and after (*right*). The original shape and dimensions of the palate were distorted by matrix that had infilled the cracks in the bone and by the tilting of various fragments, such as the plate of bone containing right molars 2 and 3. (Drawing by Dave Bichell.)

duvai as a complete or nearly complete skull for some weeks or months. Tracks on the ground and the many freshly broken fragments revealed that a herd of Masai cattle had trampled the skull, breaking it into small fragments, some of which were carried in the mud on the cows' hooves to the other side of the gorge (Leakey, 1971; Leakey and Leakey, 1964). It was not possible to recover all of the pieces of the skull. What has been found—the greater part of the occipital, two parietals, parts of both temporals, a fragment of the frontal, the left half of the palate, and isolated fragments of cranial vault—suggest strongly that the skull might have been recovered intact if it had been discovered weeks or even days earlier.

Although potentially destructive to fossils, erosion is also extraordinarily useful to the paleontologist. Fossils are sufficiently rare and unevenly distributed through the world's sedimentary rocks that it is unproductive to excavate unless an area is known to be fossiliferous. This knowledge comes from finding fossils that have been naturally excavated by erosional processes. For this reason, the density of the human population—and their ability to recognize fossils—correlates highly with the probability of discovering fossils. A few examples of how fossil assemblages have been found will illustrate this point. Many of Louis Leakey's spectacular fossil sites in East Africa were accidentally discovered by settlers who knew him or his family. Olduvai Gorge, perhaps one of Leakey's most famous sites, was found by a butterfly hunter; the Fort Ternan site was originally part of a white settler's farm, who noticed fossils eroding out of the hillside and sent them to Leakey at the National Museum. Other fossil sites, including one of the australopithecine cave sites in South Africa, were discovered by children, who took the fossils to a science teacher or museum. Some of the other South African cave sites were discovered to be fossiliferous by workmen who were quarrying limestone from the caves. Similarly, G. G. Simpson (1946) reported the discovery of an enormous deposit of fossil peccary bones by workmen clearing the back of a natural limestone cave in St. Louis that had been used as a brewery in the 1890s. Road cuts, foundation excavations, and other such work have all revealed fossil deposits.

Postdiscovery Distortion

Even after fossils have been discovered, they are not freed from the processes that distort the picture of the past that they provide. The mode of collection or excavation per se may seriously alter the characteristics of a fossil assemblage.

In the past, many paleontologists paid untrained workers to find fossils; the rate of pay was proportional to the number of pieces recovered. This practice has now been largely abandoned and is illegal in many parts of the world because it led to severe distortion of the fossil record. It does not take long for an untrained collector, whose only interest is in payment, to learn to break complete specimens into fragments and thus increase his rate of pay. In addition, such collectors cannot be expected to observe or record detailed information about the exact location of the site, stratigraphic levels, orientations, or associated material. As such information has become increasingly important in assessing fossil assemblages, so has the need for trained personnel and carefully planned excavating or collecting techniques.

COLLECTING TECHNIQUES

It is not possible to prescribe a single technique for excavating all fossil sites. In many instances, technique must be dictated by the qualities of the matrix in which the fossils are found. For example, the extraordinarily hard breccia enclosing the fossils from the South African australopithecine caves makes it impossible to use hand tools. Special machines (Figure 7.9) are commonly used to split blocks of breccia, which can then be placed in a mild acid solution if bones are present. At other sites, slow excavation using dental picks and other fine implements is appropriate. Finally, techniques may have to be adjusted according to the funding and manpower available for excavation or collection.

A marked improvement in the recovery of small bones results if the sediments can be finely sieved. Payne (1972) sieved rubble from an archaeological site to determine the number and characteristics of bones that had been overlooked during excavation. He found that standard excavating techniques recovered only 47.5 percent of the total remains of cattle, 10.1 percent of the pig remains, and 4.5 percent of the remains of sheep and goats. The recovery of very small animals, such as rodents or lizards, is improved even more dramatically if sieving is employed.

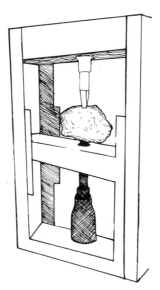

Figure 7.9 A machine used to split blocks of breccia from the South African australopithecine caves.

One of the few sites that can serve as an adequate test of the efficacy of sieving is Fort Ternan. Between 1959 and 1967, very little sieving was done, and 3,314 identifiable specimens were recovered, along with some 6,000 unidentifiable bones. In 1974 an additional 377 unidentifiable and 946 identifiable vertebrate specimens were recovered by sieving combined with standard excavating techniques.

Table 7.1 compares the faunal representation from these two excavations of the same site (see also Shipman, 1977). The percentage of very small animals recovered from the early excavation was less than 4 percent, as compared to 13 percent in the later excavation. Despite the fact that the earlier excavation included more than nine times as many specimens, the recovery of small species was only one-third as good. Clearly, sieving was responsible for this difference. Nearly twice as many genera and species were collected in the earlier excavations as in the later ones (57 versus 29). However, while the ratio of taxa is nearly 2:1, the ratio of the total

Table 7.1 Faunal representation from two excavations of Fort Ternan.

Taxon	1974		1959–1967	
	Number	Percent	Number	Percent
Birds	* 8[a]	0.1	4	0.1
Bovids	239	60.0	2,516	75.0
Carnivores	13	3.2	96	2.8
Chiroptera	* 1	0.3	* 0	0.0
Giraffoids	21	5.4	221	6.5
Insectivores	* 2	0.6	* 6	0.2
Primates	3	0.9	36	0.8
Proboscideans	16	4.1	100	3.0
Reptiles	* 12	3.2	* 1	0.02
Rhinocerotids	22	5.5	97	2.9
Rodents	* 35	9.0	* 108	3.2
Suids	1	.3	9	.3
Tragulids	0	0.0	42	1.9
Tubulidents	3	0.8	5	0.2
Indeterminate ungulates	1	0.3	73	2.2
Total	377		3,314	

a. Asterisks indicate very small animals.

number of specimens is nearly 10:1. Thus the number of taxa recovered per number of specimens is much lower in the earlier excavations. Apparently, there is a trade-off between the size of the assemblage and the rigor of the collecting technique, so that a small, rigorously collected assemblage may provide as adequate a sampling of the preserved total as a much larger assemblage collected with more standard techniques.

A different approach to evaluating excavating techniques was taken by Wolff (1975). He investigated the adequacy of samples of a fossiliferous deposit produced by first surface-collecting an area and then sieving bulk samples (45 kilograms each) of the sediments. The incremental gain in the number of genera recovered in each successive sieved sample was plotted. At one locality, the maximum number of genera (twelve) was recovered in the first seven samples; no new genera were found in the subsequent six samples (Figure 7.10). But if the remains from that locality are

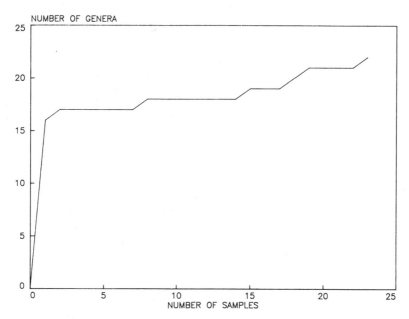

Figure 7.10 Graph showing cumulative numbers of new genera obtained with each bulk sample of sediments sieved from two Pleistocene sites in California. (From Wolff, 1975; reprinted by permission of *Paleobiology*.)

Postfossilization Distortion / 187

pooled with those from another nearby locality of the same age that probably sampled the same original animal community, new genera were still being discovered in the twenty-third sample. The pattern of increment in the number of genera follows common sense in that common taxa (those represented by more individuals and more specimens) are usually recovered before rare taxa. Also the bulk of the taxa are recovered in the earliest samples, so that the curve slopes steeply at first and then levels off.

Adding surface-collected taxa to the sieved sample makes the curve level off more rapidly. As Wolff pointed out, surface-collected material is biased in two ways. First, nearly all of the specimens represent large genera, because smaller bones and teeth are probably either washed away or buried. Second, the natural processes of weathering and erosion concentrate the remains of large animals into a lag deposit formed from an extremely large bulk sample. These processes effectively sieve the sediments for large skeletal elements. Wolff found that 72 percent of all genera recovered by all techniques were found in surface collections as well, despite the fact that large animals are usually more sparsely distributed in life than small ones. Also in life predators, especially large ones, must be substantially less densely distributed than either large or small herbivores. Wolff suggested that the ratios in a living animal community might be on the order of 10,000:100:50:1 (small herbivore:large herbivore:small predator:large predator). Wolff's samples, however, showed a ratio of 14 small herbivores to 1.5 large herbivores to 1 small predator to "a trace" (none recovered in his excavations, but some specimens previously known from the locality) of large predators. One reason that fossil assemblages do not show such large differences in the relative abundance of these trophic groups may be the bias against small animals in the fossil record. Another may be the preferential collection by researchers of rarer species.

In the same study Wolff delineated minimum sample sizes that he felt would adequately document a fauna. He concluded that 5,000–10,000 kilograms of bulk samples would yield at least 12,000–25,000 mammalian species, which will include all or nearly all of the preserved genera. From a sample of about 500 identified specimens—or about 25 specimens and 3 individuals per taxon—the relative abundance of the common mammalian taxa can be determined. Of course, generalizing from these figures assumes that the sites he sampled are roughly representative of other fossiliferous sediments.

Discoverer bias also influences the accuracy of faunal representation. Perhaps because of the difficulties involved in discovering and excavating a site, many scientists place fossils they have discovered in new taxa, resulting in a proliferation of new names for already known species. Aside from the confusion engendered by such practices, such taxonomic splitting masks true similarities between faunas. Equally confusing is the practice of lumping into one taxon specimens that exhibit extraordinary variability. The conflict between taxonomic "splitters" and "lumpers" is reminiscent of Aristophanes' dichotomy between the fox, who knows many things, and the hedgehog, who knows but one thing. The issue is not whether the fox or the hedgehog is more clever, nor is it whether species more often show extreme variability instead of speciating. The point is, rather, which style of classification (or which mode of knowing) is more useful in a particular instance as an approximation of reality.

Simpson (1963:7) pointed out the importance of thinking in terms of variable populations when making taxonomic assessments:

> Evidence that the definition (of a species) is met is largely morphological in most cases, especially for fossils. The most widely available and acceptable evidence is demonstration of a sufficient level of statistical confidence that a discontinuity exists *not* between the specimens in hand but *between the populations inferred from those specimens*. The import of such evidence and the semantic implications of the word "species" are that populations placed in separate species are either
> 1) in separate lineages (contemporaneous or not) between which significant interbreeding does not occur, or
> 2) at successive stages in one lineage but with intervening evolutionary change of such magnitude that populations differ about as much as do contemporaneous species.

8 Some Unanswered Questions

Perhaps the most painful, and exhilarating, exercise for any scientist to perform is summarizing the types of questions that can be answered, given sufficient evidence, and those that cannot. A great deal falls within the realm of the knowable, but beyond it stretches a vast expanse of the as-yet-unknowable. However, the process of delineating the two realms is valuable in that it defines the areas that need work.

With adequate evidence, it is possible to deduce a great deal about the circumstances and conditions of fossilization of any particular assemblage. With luck and skill, one can draw many conclusions about the animal communities from which the assemblage was derived and the paleoenvironment in which they lived. Geologic evidence may permit reconstruction of the local and regional geomorphology and conditions of deposition. The latter may also be based on analysis of skeletal representation in the fossil assemblage. Information such as the presence and type of zeolites, mudcracks, or caliche deposits may help establish the local climate, as can the ratios of different oxygen isotopes. From these types of information, a large-scale reconstruction can be made, and details about specific habitats filled in by the evidence of preserved plant parts, analysis of fossil pollens, and anatomical or taxonomic studies of members of the local animal community.

But delineating the animal community or communities requires other types of evidence. The relative abundance and representation of different species can be calculated, and their physical (anatomical) adaptations can be compared with those of living species of known habitat preference. Predator/prey ratios in the original population can be roughly approximated by using the MNI of different species, although this may not be accurate if the modes of death of predators and prey have caused different distortions of their original numbers. A general impression of predator/prey ratios can be gleaned from the number of tooth marks and other signs of predator damage on the fossils.

One can identify what species lived together from evidence of a common taphonomic history. Data on the spatial distribution, state of preservation, skeletal representation, abundance of species, and size of the animals are useful in deducing the taphonomic histories of specimens, species, or whole assemblages. Two aspects of the taphonomic history that are particularly important are the agent of collection and the agents of breakage. Along with the data mentioned above, the size of the bones and the sedimentary evidence will help identify the agent of collection. In at least some instances, the agent of breakage can be identified by comparing breakage patterns to those of known cause. Closely tied to these aspects of taphonomic history are the mode of death and the mortality pattern. Although the cause of death can only rarely be identified, attritional and catastrophic mortality can be distinguished by the age structure of the preserved population. If the mortality pattern is catastrophic, realistic population estimates can be made. In some instances, the season of death and of deposition can be established. Lakebeds, for example, often show cyclical deposition of sediments; studies of population structure and age classes may also help pinpoint seasonal deaths.

Taphonomic and paleoecological studies make it possible to extend our knowledge of the past beyond the study of single organisms or species. It is possible to place a species in its appropriate environmental context and habitat and to identify the other plants and animals that lived with it. In special circumstances food chains can be reconstructed in whole or part. Direct evidence of diet is occasionally available in the form of coprolites, raptor pellets, and preserved stomach contents. At hominid sites, there may be signs of hearths and charred or cut bones. Visible marks of predator activity, such as characteristic canine tooth punctures or gnawing marks, also help in reconstructing food chains, as do the extremely rare instances in which a predator and its prey are preserved in a death pose. Also useful are analysis of the strontium content of fossils and inspection of microscopic wear on teeth. This is an impressive amount of information about the past, and our ability to extract this and more information is rapidly expanding.

But what about the unknowns? The lives and habits of even living species exhibit so many details and complexities that it is unrealistic to think we will ever be able to trace such details in the lives of all past species. Some of the more elusive and intriguing questions that we cannot yet answer are discussed below.

Did all the individuals preserved in one place habitually live together? Even in cases of catastrophic mortality, it is difficult to rule out the possibility that what is preserved is not a normal community of individuals. In the case of attritional mortality, there is at present no means of determining whether or not the animals ever saw one another in life. And in all but exceptional cases, we can't tell what the animals looked like in life, either, because soft tissues are so rarely preserved. Artist's reconstructions tend to fulfill our prejudices of what a particular animal's ancestors "ought" to have looked like, but in reality we have no evidence of whether a species was furred or smooth skinned, striped, spotted, or plain in color (Figure 8.1).

Studies of living species reveal that animals possess remarkable plasticity in their locomotor, feeding, and social behavior. For example, even species showing marked adaptations to particular diets or locomotor patterns regularly vary their behavior in response to minor habitat differences or, apparently, to whim. Thus we can look for major functional adaptations to locomotor patterns, to diets dominated by one type of food, and to climate or habitat through social behavior or physiology. However, even if we can identify such complexes of adaptations, we must assume that the animals of the past were probably capable of considerable but unmeasurable plasticity in behavior. Much of this variability and plasticity concerns the more subtle aspects of a niche, which is

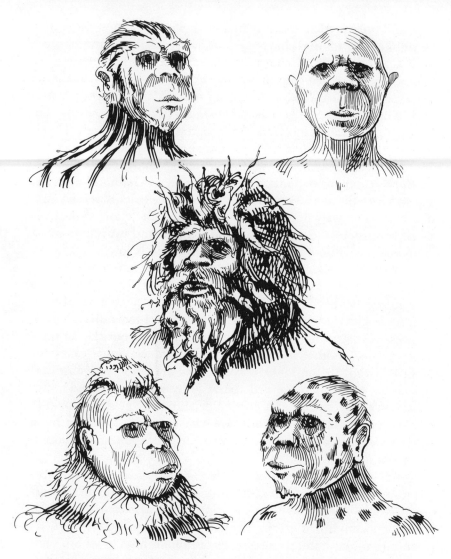

Figure 8.1 Various reconstructions of the soft tissues of *Australopithecus* change its appearance, even though all are based on the same skull. (Drawing by Dave Bichell.)

why delineating a niche is a difficult task requiring years of patient study of living species. At present, paleoecologists are simply unable to define the niche of an extinct species with any precision. A major recent breakthrough is the discovery that an animal's pre-

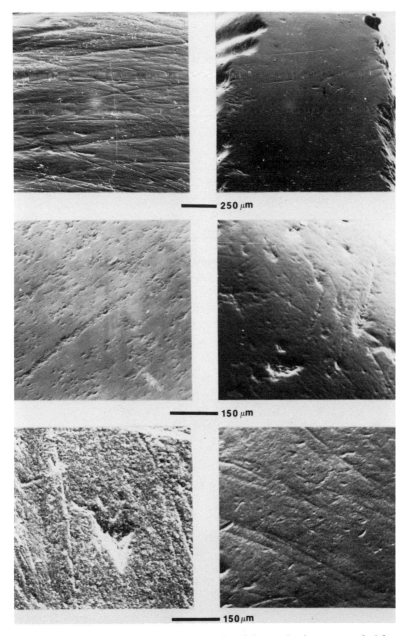

Figure 8.2 Scanning electron micrographs of the teeth of species with different diets. *Top left*, white rhinoceros (grazer); *top right*, giraffe (browser); *middle left*, cheetah (meat-eater); *middle right*, orangutan (frugivore); *bottom left*, spotted hyena (bone-crusher); and *bottom right*, robust australopithecine (frugivore?). (Photo courtesy of Alan Walker.)

Some Unanswered Questions / 195

dominant diet, broadly defined, in the last few months before death can be determined by inspecting the wear on its teeth with a scanning electron microscope (Walker, Hoeck, and Perez, 1978; Walker, 1979 and in press); see Figure 8.2. Although such work provides invaluable evidence about ancient diets, it is still far from an accurate representation of the range and variability of the diet of any individual.

One insoluble dilemma about the past concerns unique anatomical adaptations. In the absence of living analogues, we have no way of knowing the functions of such adaptations, because of the inadequacy of our understanding of biomechanics and functional anatomy. Taphonomic and paleoecological studies may link a species firmly with one particular habitat, but our knowledge is too limited to tell us in detail what the species was doing in that habitat unless we have modern analogues to draw upon.

It is also nearly impossible to answer the question, "How did this animal die?" Forensic medicine is an old and complex science, but it was developed to deal with whole bodies, not fragmentary skeletal remains. Few modes of death leave unambiguous osteological evidence, and fewer still leave marks that survive fossilization and fragmentation.

Finally, our knowledge of the past is always inadequate to determine the true set of evolutionary relationships among extinct species and the forces that caused the divergences we see. It is important to realize that phylogenies are constructs imposed upon faulty and incomplete data; this is why they are continually under revision and are the subject of controversy. Lowenstein (1979) has suggested that the remnants of collagen that survive in fossils may be species-specific and that immunologic reactions may be used to identify species with certainty once the techniques are sufficiently refined. But even these techniques cannot reveal the driving forces or the twisting paths of evolution.

These are but a few of the questions to be answered, some—perhaps all—of which will be resolved by clever researchers using innovative and imaginative techniques. But as surely as questions are answered, new, more complex ones will arise. I hope that in years to come we will see the past less dimly, until we are ready to answer the ultimate question: "What was it really like?"

GLOSSARY

REFERENCES

INDEX

Glossary

Agent An animal, person, or force that acts upon a bone assemblage, as an agent of collection or agent of damage.
Allochthonous Derived from some other place or habitat.
Alluvium A general term for fluvial or waterlaid sediments.
Assemblage A collection of modern or fossil bones in one place.
Autochthonous Locally derived; untransported.

Biocoenose A group of organisms that live together in one place; a life assemblage. (See *Taphocoenose; Thanatocoenose.*)
Biomass The weight of living organisms in an area.
Bioturbation The disturbance of sediments by living organisms, as by trampling or burrowing.

Channel The physical structure cut by a stream or river and through which it flows.
Chopping mark A V-shaped groove made by striking a bone with a stone tool at an angle roughly perpendicular to the surface of the bone, characterized by fragments of bone crushed inward at the bottom of the groove. (See Figure 5.4.)
Community A complex of plants and animals living in one place, bound by a network of interdependencies and having a characteristic trophic structure.
Correlation coefficient A statistical measure of the degree to which two variables covary; also known as Pearson's *r*.

Cracking, mosaic   A weathering stage in which subchondral bone on articular surfaces breaks into many irregular pieces. (See Figure 7.5.)

Cracking, polygonal   A pattern seen on subchondral bone on articular surfaces produced by burning; the bone breaks into many concave, polygonal plates with vascular canals at the corners (See Figure 7.5.)

Critical shear stress   The velocity necessary to initiate movement of a given particle. Synonym: *Threshold drag.* (See formula 2.9.)

CSI   Corrected Number of Specimens per Individual. The number of specimens divided by the number of individuals in an assemblage and corrected for the number of bones in a whole skeleton of that species. (See formula 6.2.)

Cutting mark   Any mark made on a bone by a tool made of stone, metal, or other material. (See *Slicing mark, Chopping mark.*)

Deformation, plastic   A postdepositional and sometimes postfossilization change in the shape of a fossil or bone without breakage, usually caused by compaction from the weight of overlying sediments.

Delta   The structure created by the deposition of riverborne sediments at the river's mouth. (See Figure 3.4.)

Density   The weight of an object divided by its volume.

Deposition   Burial in sediments.

Diagram, rose   A means of representing orientation graphically by radially arranged wedges that show the number of bones at various orientations between 0 and 360 degrees. A mirror-image rose diagram sums orientation data in wedges opposite (at 180 degrees) to each other. (See Figure 4.5.)

Diameter, nominal   The diameter of a quartz sphere having the same volume as a bone or other nonspherical particle; nominal diameter is used to predict settling velocity. (See formulas 2.7 and 2.8.)

Dip   The angle of declination of an object or plane.

DSE   Distinctive Skeletal Element. A skeletal element that is readily recognizable even when fragmentary because of its distinctive texture, markings, or shape.

Fauna   A group of animal species that presumably lived together, found as fossils in one place.

Fluvial   Of or pertaining to a river.

Fracture, columnar   A bone break forming many rectangular or nearly rectangular pieces, commonly found in dried bones. (See Figure 7.1.)

Fracture, perpendicular smooth   A break with a flat, smooth fracture surface that is roughly perpendicular to the long axis of the bone; a typical postfossilization break. (See Figure 7.3.)

Fracture, spiral A break in which the fracture surface curves around the shaft of a long bone. A type I spiral fracture follows the predominant direction of the collagen fibers of the bone; a type II spiral fracture cuts across that direction. (See Figure 7.2.)

FRI Faunal Resemblance Index. An indication of the similarity of two faunas, based on the number of species in common and the total number of species in the smaller fauna. (See formula 6.3.)

Horizon A plane within a stratum or geologic bed.

Hydraulic Of or pertaining to water.

Hydrodynamic Of or pertaining to moving water.

Igneous Formed by the solidification of molten magma; one of the three basic types of rock. (See also *Metamorphic; Sedimentary.*)

Impregnation The addition of minerals to a bone during fossilization. Synonym: *Permineralization.*

Kolmogorov-Smirnov test A statistical means of assessing the similarity of two sets of data arranged by ranks or classes. (See formula 6.5.)

K-selection A reproductive pattern in which a species has few offspring that have long periods of dependency and sexual immaturity, typical of species that live in social groups and have relatively long lifespans. (see *r-selection.*)

Lacustrine Of or pertaining to a lake.

Lag deposit An accumulation of bones or other sedimentary particles that are too dense to be moved by the existing wind or water current. (See *Voorhies Group.*)

Life table A standardized means of showing the age distribution and mortality rates of different age classes in a population. (See Table 6.5.)

Living floor A surface preserving the remains of a site inhabited by hominids.

Metamorphic Formed by chemical or physical alteration of another type of rock at high temperature or pressure; one of the three basic types of rock. (See also *Igneous; Sedimentary.*)

MNI Minimum Number of Individuals. The smallest number of individuals of one species from which the most common skeletal element in an assemblage could have been derived.

Mortality, attritional Death from natural causes at a normal rate in a population. (See Figure 6.15.)

Mortality, catastrophic Death of all or most of a population simultaneously as the result of some catastrophe, such as a flood or epidemic disease. (See Figure 6.15.)

Neotaphonomy The branch of taphonomy that studies the modern processes of death, decay, destruction, dispersal, and concentration of skeletal remains.

NSI Number of Specimens per Individual. The number of bones attributed to a taxon divided by the MNI of that taxon. (See formula 6.1.)

Orientation The azimuth, or compass direction, of the long axis of a bone, as measured on a horizontal plane.

Overbank deposit A quantity of fine-grained, fluvial sediments left on a floodplain by a river that has overflowed its banks. (See Figure 3.2.)

Oxbow lake A lake formed by a loop of a meandering river that is cut off from the current; typical sedimentary deposits are fine-grained. (See Figure 3.2.)

Paleoecology The study of ancient ecosystems and the habits, lifestyles, and interdependencies of the species in that ecosystem.

Paleomagnetic dating A method for determining the age of rocks that depends upon (1) the orientation of magnetic particles in those rocks, and (2) the reversals of the north and south poles at various times in the earth's history.

Paleotaphonomy The branch of taphonomy that studies the characteristics of fossil assemblages and their enclosing sediments.

Pearson's r A correlation coefficient; a statistical measure of the degree to which two variables covary. Synonym: *Correlation coefficient.*

Percentage representation The number of specimens of a skeletal element in an assemblage divided by the number expected on the basis of the MNI for that species. Synonym: *Survival rate.*

Permineralization The addition of minerals to a bone during fossilization. Synonym: *Impregnation.*

Phreatic zone The level below the land surface that is just below the water table. Synonym: zone of saturation. (See Figure 3.7 and *Vadose zone.*)

Point bar A deposit of sediments that builds up on the inner, convex curve of a meandering river. (See Figure 3.2.)

Postdepositional Of or pertaining to the period between the burial of bones in sediments and their fossilization.

Postfossilization Of or pertaining to the period after fossilization.

Postmortem Of or pertaining to the period between the death of an animal and burial of its bones in sediments.

Predation curve A graphic representation of the rates of predation upon individuals in different age classes of one species; the more intense the predation, the steeper the slope. (See Figures 6.18 and 6.19.)

Preservation potential The likelihood of preservation of a particular skeletal element, based on its physical attributes and taphonomic history.

Prodelta The leading edge of an outward-building delta, characterized by fine-grained deposits. (See Figure 3.2.)

Puncture A roughly circular hole or depressed fracture in a bone, made by a carnivore's canine. (See Figure 5.4.)

RAI Relative Abundance Index. An indication of the similarity of two faunas, based on the number of species in common and the number of individuals in each fauna. (See formula 6.4.)

r-selection A reproductive pattern in which a species has many offspring that mature rapidly and receive little parental care, typical of species that have short lifespans and little capacity for learned behavior. (See *K-behavior.*)

Sampling, X-type A fossil assemblage that is a relatively undistorted representation of the original community, because the taphonomic history has destroyed bones without regard to ecological indicators. (See Figure 6.11 and *Sampling, Y-type.*)

Sampling, Y-type A fossil assemblage that is a distorted representation of the original community, because the taphonomic history has destroyed bones of animals with similar ecological habits.

SA/V ratio Surface Area to Volume ratio. A measure of the deviation of a shape from spherical and, therefore, an indicator of the inaccuracy of predicted hydraulic behavior.

S/C ratio Spongy to Compact bone ratio. A measure of bone density based on the proportions of spongy (cancellous) and compact bone. (See Figure 2.2.)

Sedimentary Of or pertaining to particles (sediments) that have settled out of the air or water; one of three basic types of rocks and the one in which fossils are most commonly found. (See *Igneous; Metamorphic.*)

Settling velocity The speed at which a particle drops through a column of standing water; a basic measure of hydraulic behavior. (See formula 2.6.)

SI Shape Index. The maximum length of a bone divided by its maximum breadth, a gross indicator of the shape of a bone. (See formula 2.5.)

Slicing mark A V-shaped groove made by drawing a metal or stone edge along a surface in the direction of the long axis; characterized by multiple fine striations within the main groove. (See Figure 5.3.)

Square frequency A measure of the abundance of a species or skeletal element based on the number of standard-size squares in which the species or element is present, divided by the total number of squares sampled.

Strain ellipse The ellipse formed by deforming a circle in the same direction and to the same extent that a fossil has been deformed; used in mathematically reconstructing the original shape of the fossil.

Strain, homogeneous A type of deformation in which the only distortion is a uniform lengthening or shortening of all lines or areas of a fossil in one direction.

Stratum A geologic bed.

Strike An axis or line on a plane that remains horizontal as the plane dips.

Survivorship curve A graphic representation of the number of living individuals in each age class in a population. (See Figure 6.13.)

Taphocoenose An assemblage of bones or fossils brought together by taphonomic processes after the death of the animals. (See *Biocoenose; Thanatocoenose.*)

Taphonomic history The catalogue of postmorten, postdepositional, and postfossilization events that have influenced the preservation of an assemblage; the emphasis is often on postmortem events.

Taphonomy The study of the transition of skeletal elements from parts of living animals to fossilized fragments; also, the processes or events affecting bone destruction or preservation during this transition. (See Figure 1.6.)

Thanatocoenose A death assemblage or assemblage of bones of animals that died in the same place. (See *Biocoenose; Taphocoenose.*)

Tooth mark A groove made by an animal in chewing on the bone, characterized as a round-bottomed groove without striations. (See Figure 5.3.)

Trophic Of or pertaining to feeding or diet.

Trophic replacement The substitution, over time, of one species for another having a similar trophic role in the community. (See Table 6.2.)

Tuff A stratum formed by consolidation of volcanic ash; frequently used in radiometric dating.

T/V ratio Teeth to Vertebrae ratio. The number of teeth in an assemblage divided by the number of vertebrae; a measure of the degree of sorting of the assemblage.

Vadose zone The level below the land surface that is just above the water table. Synonym: zone of aeration. (See Figure 3.7 and *Phreatic zone*.)

Voorhies Group A grouping of skeletal elements on the basis of their similar potential for hydraulic dispersal. (See Table 2.1.)

References

Ager, D. V., 1963. *Principles of Paleoecology.* McGraw-Hill, New York.

Allan, J. R. L., 1970. *Physical Processes of Sedimentation.* Allen and Unwin, London.

Andrews, P., and E. Nesbit Evans, 1979. "The environment of *Ramapithecus* in Africa." *Paleobiol.* 5(1):22–30.

Andrews, P., and A. Walker, 1976. "The primate and other fauna from Fort Ternan, Kenya." In Glynn Ll. Isaac and Elizabeth R. McCown, eds., *Human Origins: Louis Leakey and the East African Evidence.* W. A. Benjamin, Menlo Park, Calif.

Andrews, P., and J. A. H. Van Couvering, 1975. "Palaeoenvironments in the East African Miocene." In F. Szaley, ed., *Approaches to Primate Paleobiology.* Vol. 5, Contrib. Primatol.

Andrews, P., J. M. Lord, and E. Nesbit Evans, 1979. "Patterns of ecological diversity in fossil and modern mammalian faunas." *Biol. J. Linnaean Soc.* 11:177–205.

Andrews, P., J. Van Couvering, and J. Van Couvering, 1972. "Rusinga 1971 Expedition—Associations and Environment in the Miocene of Kaswanga Point." Unpublished report.

Ardrey, R., 1976. *The Hunting Hypothesis: A Personal Conclusion Concerning the Evolutionary Nature of Man.* Atheneum, New York.

Ardrey, R., 1966. *The Territorial Imperative.* Atheneum, New York.

Ardrey, R., 1961. *African Genesis: A Personal Investigation in the Animal Origins and Nature of Man.* Dell, New York.

Bearder, S. K., 1977. "Feeding habits of spotted hyaenas in a woodland habitat." *E. Afr. Wildl. J.* 15:263–280.

Behrensmeyer, A. K., 1978. "Taphonomic and ecologic information from bone weathering." *Paleobiol.* 4(2):150–162.

Behrensmeyer, A. K., 1976. "Fossil assemblages in relation to sedimentary environments." In Y. Coppens, F. C. Howell, G. Ll. Isaac, and R. E. F. Leakey, eds., *Earliest Man and Environments in the Lake Rudolf Basin: Stratigraphy, Paleoecology and Evolution.* University of Chicago Press, Chicago.

Behrensmeyer, A. K., 1975. "Taphonomy and paleoecology of the Plio-Pleistocene vertebrate assemblages east of Lake Rudolf, Kenya." *Mus. Comp. Zool. Bull.* 146:473–578.

Behrensmeyer, A. K., 1973. "The taphonomy and paleoecology of the Plio-Pleistocene vertebrate assemblages east of Lake Rudolf, Kenya." Unpublished Ph.D. thesis, Harvard University, Cambridge, Mass.

Behrensmeyer, A. K., and D. Dechant, 1980. "The recent bones of Amboseli Park, Kenya, in relation to East African paleoecology." In A. K. Behrensmeyer and A. Hill, eds., *Fossils in the Making.* University of Chicago Press, Chicago.

Behrensmeyer, A. K., D. Western, and D. Dechant-Boaz, 1979. "New perspectives in vertebrate paleoecology from a recent bone assemblage." *Paleobiol.* 5(1):12–21.

Berry, L. G., and B. Mason, 1959. *Mineralogy.* W. H. Freeman, San Francisco.

Binford, L. R., and J. B. Bertram, 1977. "Bone frequencies and attritional processes." In L. R. Binford, ed., *For Theory Building in Archaeology.* Academic Press, New York.

Bishop, W. W., 1980. "Paleogeomorphology and continental taphonomy." In A. K. Behrensmeyer and A. Hill, eds., *Fossils in the Making.* University of Chicago Press, Chicago.

Bishop, W. W., 1976. "Thoughts on the workshop: stratigraphy, paleoecology and evolution in the Lake Rudolf Basin." In Y. Coppens, F. C. Howell, G. Ll. Isaac, and R. E. F. Leakey, eds., *Earliest Man and Environments in The Lake Rudolf, Basin: Stratigraphy, Paleoecology and Evolution.* University of Chicago Press, Chicago.

Bishop, W. W., 1963. "The later Tertiary and Pleistocene in eastern equatorial Africa." In F. C. Howell and F. Bourlière, eds., *African Ecology and Human Evolution.* Aldine, Chicago.

Bishop, W. W., and G. R. Chapman, 1970. "Early Pliocene sediments and fossils from the Northern Kenya Rift Valley." *Nature* (London) 226:914–918.

Bishop, W. W., J. A. Miller, and F. J. Fitch, 1969. "New potassium-argon age determinations relevant to the Miocene fossil mammal sequence in East Africa." *Am. J. Sci.* 267:669–699.

Boaz, N. T. and A. K. Behrensmeyer, 1976. "Hominid taphonomy: Transport of human skeletal parts in an artificial fluviatile environment." *Am. J. Phys. Anthrop.* 45:53–60.

Boaz, N. T., and J. Hampel, 1978. "Strontium content of fossil tooth enamel and diet of early hominids." *J. Paleontol.* 52(4):929–933.

Bonnichsen, R., 1979. "Pleistocene Bone Technology in the Beringian Refugium." *Arch. Survey Canada*, paper no. 89.

Bonnichsen, R., 1978. "Critical arguments for Pleistocene artifacts from the Old Crow Basin, Yukon: a preliminary statement." In A. L. Bryan, ed., *Early Man in America from a Circum-Pacific Perspective*. Occasional papers no. 1, Dept. of Anthropology, University of Alberta. Archaeological Researches International, Edmonton.

Bourlière, F., 1973. "The comparative ecology of rain forest mammals in Africa and tropical America: some introductory remarks." In B. J. Meggers, E. S. Ayensu, and W. D. Duckworth, eds., *Tropical Forest Ecosystems in Africa and South America: A Comparative Review*. Smithsonian, Washington, D.C.

Bourlière, F., 1963. "Observations of the ecology of some larger African mammals." In F. C. Howell and F. Bourlière, eds., *African Ecology and Human Evolution*. Aldine, Chicago.

Brain, C. K., 1980. "Some criteria for the recognition of bone-collecting agencies in African caves." In A. K. Behrensmeyer and A. Hill, eds., *Fossils in the Making*. University of Chicago Press, Chicago.

Brain, C. K., 1978. "Some aspects of the South African australopithecine sites and their bone accumulations." In C. J. Jolly, ed., *African Hominidae of the Plio-Pleistocene*. Duckworth, London.

Brain, C. K., 1976. "Some principles in the interpretation of bone accumulations associated with man." In G. Ll. Isaac and E. McCown, eds., *Human Origins*. W. A. Benjamin, Menlo Park, Calif.

Brain, C. K., 1974. "Some suggested procedures in the analysis of bone accumulations from southern African Quaternary sites." *Ann. Trans. Mus.* 29(1):1–8.

Brain, C. K., 1970. "New finds at the Swartkrans australopithecine site." *Nature* 225:1112–1119.

Brain, C. K., 1967. "Hottentot food remains and their bearing on the interpretation of fossil bone assemblages." *Sci. Pap. Nam. Des. Res. Sta.* 32:1–11.

Brain, C. K., 1958. "The Transvaal ape-man-bearing cave deposits." *Mem. Trans. Mus.* 11:1–131.

Butzer, K. W., 1976. *Geomorphology from the Earth*. Harper and Row, New York.

Butzer, K. W., 1971. "The lower Omo basin: geology, fauna and hominids of Plio-Pleistocene formations." *Naturwissenschaften* 58:7–16.

Cannon-Bonventre, K., E. Engelman, D. Kent, and P. Shipman, 1977. *A Literature Study to Evaluate Health Parameters in Various Human Populations in Relation to Diet*. U.S. Government Printing Office, Washington, D.C.

Clark, J., and T. E. Guensberg, 1970. "Population dynamics of *Leptomeryx*." *Field. Geol.* 16:411–451.

Clark, J., J. R. Beerbower, and K. K. Kietzke, 1967. "Oligocene sedimentation, stratigraphy and paleoecology and paleoclimatology in the Big Badlands of South Dakota." *Field Geol. Mem.* 5:1–158.

Coe, M., 1980. "The role of modern ecological studies in the reconstruction of paleoenvironments in sub-Saharan Africa." In A. K. Behrensmeyer and A. Hill, eds., *Fossils in the Making*. University of Chicago Press, Chicago.

Coe, M. D., and K. Flannery, 1964. "Microenvironments and Mesoamerican prehistory." *Science* 143:650–654.

Crader, D., 1974. "The effects of scavengers on bone material from a large mammal: an experiment conducted among the Bisa of the Luangwa Vallèy, Zambia." Monograph 4, *Archaeological Survey, Inst. of Archaeology*, University of California, Los Angeles.

Dart, R. A., 1959. *Adventures with the Missing Link*. Institutes Press, Philadelphia.

Dart, R. A., 1957. "The osteodontokeratic culture of *Australopithecus prometheus*." *Trans. Mus. Mem.* no. 10.

Dart, R. A., 1949. "The predatory implemental technique of *Australopithecus*." *Amer. J. Phys. Anthrop.* 7:1–38.

Davis, D. H. S., 1959. "The barn owls' contribution to ecology and paleoecology." *Proc. 1st Pan-African Ornith. Cong., Ostrich Supple.* 3:144–153.

Deevey, E. S., 1947. "Life tables for natural populations of animals." *Quart. Rev. Biol.* 22:283–314.

de Graaf, G., 1960. "A preliminary investigation of the mammalian microfauna in Pleistocene deposits of caves in the Transvaal system." *Palaeont. Afr.* 7:59–118.

Dodson, P., 1973. "The significance of small bones in paleoecological interpretation." *Contrib. Geol.* 12:15–19.

Dodson, P., and Wexlar, D., 1979. "Taphonomic investigations of owl pellets." *Paleobiol.* 5:279–284.

Dorst, J., and Dandelot, P., 1970. *A Field Guide to the Larger Mammals of East Africa.* Collins, London.

Douglas-Hamilton, I., and O. Douglas-Hamilton, 1974. *Among the Elephants.* Viking Press, New York.

Efremov, J. A., 1940. "Taphonomy: new branch of paleontology." *Pan-American Geologist* 74(2):81–93.

Einarsen, A. S., 1956. "Determination of some predator species by some field signs." *Oregon State Monographs: Studies in Zoology* 10:1–34.

Evans, F. G., 1973. *Mechanical Properties of Bone.* Charles C. Thomas, Springfield, Ill.

Evernden, J. F., and G. H. Curtis, 1965. "Potassium-argon dating of late Cenozoic rocks in East Africa and Italy." *Curr. Anthrop.* 6(4):343–385.

Evernden, J. F., D. Savage, G. H. Curtis, and G. T. James, 1964. "Potassium-argon dates and the Cenozoic mammalian chronology in North America." *Am. J. Sci.* 262(2):145–198.

Ewer, R. F., 1973. *The Carnivores.* Cornell University Press, Ithaca, N.Y.

Falk, C. R., 1977. "Analyses of unmodified vertebrate fauna from sites in the middle Missouri subarea: a review." *Plains Anthrop.* 22:150–161.

Fleagle, J., 1978. "Size distributions of living and fossil primate faunas." *Paleobiol.* 4:67–76.

Frison, G. C., 1976. "Cultural activity associated with prehistoric mammoth butchering and processing." *Science* 194:728–730.

Gifford, D., 1980. "Ethnoarcheological contributions to the taphonomy of human sites." In A. K. Behrensmeyer and A. Hill, eds., *Fossils in the Making.* University of Chicago Press, Chicago.

Gifford, D., 1978. "You walk by and I fall to pieces: a longitudinal taphonomic study of large ungulates at Lake Turkana, Kenya." Paper presented to the American Association of Physical Anthropologists, Toronto.

Gifford, D., 1977. "Observation of modern human settlements as an aid to archaeological interpretation." Unpublished Ph.D. thesis, University of California, Berkeley.

Glob, P., 1954. "Lifelike man preserved 2000 years in past." *Nat. Geog.* 105:419–430.

Goodman, L. A., 1954. "Kolmogorov-Smirnov test for psychological research." *Psych. Bull.* 51:160–168.

Gray, T., 1978. "Environmental reconstruction of the Hadar formation with implications for hominid adaptations and taxonomy." Unpublished Ph.D. thesis, Case Western Reserve University.

Grayson, D. K., 1978. "Minimum numbers and sample size in vertebrate faunal analysis." *Am. Antiq.* 43:53–65.

Hanson, C. B., 1980. "Fluvial taphonomic processes: models and experiments." In A. K. Behrensmeyer and A. Hill, eds., *Fossils in the Making*. University of Chicago Press, Chicago.

Hay, R. L., 1976. *Geology of the Olduvai Gorge: A Study of Sedimentation in a Semiarid Basin*. University of California Press, Berkeley and Los Angeles.

Haynes, G., in press. "Prey bones and predators: potential ecologic information from analysis of bone sites." *Ossa* 7.

Haynes, G., 1980. "Evidence of carnivore gnawing on Pleistocene and Recent mammalian bones." *Paleobiol.* 6(3):341–351.

Hill, A., 1980. "Early postmortem damage to the remains of some contemporary East African mammals." In A. K. Behrensmeyer and A. P. Hill, eds., *Fossils in the Making*. University of Chicago Press, Chicago.

Hill, A., 1978. "Taphonomical background to fossil man: problems in paleoecology." In W. W. Bishop, ed., *Geological Background to Fossil Man*. Scottish Academic Press, University of Toronto Press, Edinburgh and Toronto.

Hill, A., 1976. "On carnivore and weathering damage to bone." *Curr. Anthrop.* 17(2):335–336.

Hill, A., 1975. "Taphonomy of contemporary and Late Cenozoic East African vertebrates." Unpublished Ph.D. thesis, University of London.

Hill, A., and A. Walker, 1972. "Procedures in vertebrate taphonomy." *J. Geol. Soc. Lond.* 128:399–406.

Johanson, D. C., M. Splinger, and N. T. Boaz, 1976. "Paleontological excavations in the Shungura formation, lower Omo Basin, 1969–1973." In Y. Coppens, F. C. Howell, G. Ll. Isaac, and R. E. F. Leakey, eds., *Earliest Man and Environments in the Lake Rudolf Basin: Stratigraphy, Paleoecology and Evolution*. University of Chicago Press, Chicago.

Johnson, R. G., 1960. "Models and methods for the analysis of the mode of formation of fossil assemblage." *Geol. Soc. Am. Bull.* 71:1075–1086.

Kay, R., and Cartmill, M., 1977. "Cranial morphology and adaptations of *Palaechthon nacimienti* and other Paromomyidae (Pleisiadapoidea, ? Primates), with a description of a new genus and species." *J. Hum. Evol.* 6:1–36.

Klein, R. G., 1978. "Stone age predation on large African bovids." *J. Arch. Sci.* 5:195–217.

Klein, R. G., 1975. "Paleoanthropological implications of the nonarcheological bone assemblage from Swartklip 1, South-Western Cape Province, S. Africa." *Quat. Res.* (University of Washington) 5:275–288.

Klein, R. G., 1969. *Man and Culture in the Late Pleistocene: A Case Study*. Chandler, San Francisco.

Korth, W. W., 1979. "Taphonomy of microvertebrate fossil assemblages." *Ann. Carnegie Mus.* 48:235–285.

Krinsley, D., and W. Wellendorf, 1980. "Wind velocities determined from the surface textures of sand grains." *Nature* (London) 283:372–373.

Krinsley, D., and J. Donahue, 1968. "Environmental interpretation of sand grain surface texture by electron microscopy." *Geol. Soc. Amer. Bull.* 1979:743–748.

Kruuk, H., 1976. "Feeding and social behavior of the striped hyaena (*Hyaena vulgaris* Desmarest)." *E. Afr. Wildl. J.* 14:91–112.

Kruuk, H., 1972. *The Spotted Hyena: A Study of Predation and Social Behavior.* University of Chicago Press, Chicago.

Kurtén, B., 1953. "On the variation and population dynamics of fossil and recent mammal populations." *Acta Zool. Fennica* 76:1–122.

Lakes, R. and S. Saha, 1979. "Cement line motion in bone." *Science* 208:501–503.

Leakey, L. S. B., 1968. "Bone smashing by late Miocene Hominidae." *Nature* 218:528–530.

Leakey, L. S. B., 1961. "A new lower Pliocene fossil primate from Kenya." *Ann. Mag. Nat. Hist.*, series 4:689–696.

Leakey, L. S. B., and M. D. Leakey, 1964. "Recent discoveries of fossil hominids in Tanganyika: at Olduvai Gorge and near Lake Natron." *Nature* (Lond.) 202:5–7.

Leakey, M. D., 1979. "Footprints in the ashes of time." *Nat. Geog.* 155:4.

Leakey, M. D., 1971. *Olduvai Gorge.* vol. 3. Cambridge University Press, London.

Leakey, M. D., and R. Hay, 1979. "Pliocene footprints in the Laetoli beds at Laetoli, northern Tanzania." *Nature* 278:317–323.

Leopold, L. B., G. S. Wolman, and J. P. Miller, 1964. *Fluvial Processes in Geomorphology.* W. H. Freeman, San Francisco.

Lowenstein, J. M., 1979. "Immunospecificity of fossil collagens." Paper presented to the American Association of Physical Anthropologists, San Francisco.

Lyman, R. L., 1979. "Available meat from faunal remains: a consideration of techniques." *Am. Antiq.* 44(3):536–546.

MacArthur, R., and E. O. Wilson, 1967. *Theory of Island Biogeography.* Princeton University Press, Princeton, N.J.

MacDonald, J. D., 1973. *Birds of Australia.* H. F. and G. Witherby, London.

Makacha, S., and G. B. Schaller, 1969. "Observations on lions in the large Manyara National Park, Tanzania." *E. Afr. Wildl. J.* 7:99–103.

Mardia, K. V., 1972. *Statistics of Directional Data.* Academic Press, New York.

Mech, D. L., 1970. *The Wolf: The Ecology and Behavior of an Endangered Species*. Natural History Press, New York.

Mech, D. L., 1966. *The Wolves of Isle Royale*. National Parks Service Fauna, no. 7.

Medawar, P., 1974. "A geometric model of reduction and emergence." In F. I. Ayala and T. Dobzhansky, eds., *Studies in the Philosophy of Biology: Reduction and Related Problems*. University of California Press, Berkeley.

Mellett, J. S., 1974. "Scatological origins of microvertebrate fossils." *Science* 185:349–350.

Miller, G. J., 1975. "A study of cuts, grooves and other marks on recent and fossil bone: II. Weathering cracks, fractures, splinters and other similar phenomena." In E. H. Swanson, ed., *Lithic Technology: Making and Using Stone Tools*. Aldine, Chicago.

Miller, G. J., 1969. "A study of cuts, grooves and other marks on recent and fossil bone: I. Animal tooth marks." *Tebiwa* 12:20–26.

Mills, M. G. L., and M. E. J. Mills, 1977. "An analysis of bones collected at hyaena breeding dens in the Gemsbok National Parks (Mammalia: Carnivora)." *Ann. Trans. Mus.* 30(14):145–155.

Mueller, H. C., 1975. "Hawks select odd prey." *Science* 188:953–954.

Murie, A., 1944. *Wolves of Mount McKinley*. National Parks Service Fauna, no. 5.

Myers, T., M. R. Voorhies, and R. G. Corner, 1980. "Spiral fractures and bone pseudotools at paleontological sites." *Am. Antiq.* 45(3):483–489.

Noe-Nygaard, N., 1977. "Butchering and marrow fracturing as a taphonomic factor in archaeological deposits." *Paleobiol.* 3:218–237.

Odum, E. P., 1971. *Fundamentals of Ecology*, 3rd ed. W. B. Saunders, Philadelphia.

Olson, E. C., 1966. "Community evolution and the origin of mammals." *Ecol.* 47:291–302.

Olson, E. C., 1962. "Late Permian terrestrial vertebrates, U.S.A. and U.S.S.R." *Trans. Amer. Phil. Soc.* 52(2):3–224.

Olson, E. C., 1961. "Food chains and the origin of mammals." *International Colloquium on the Evolution of Lower and Unspecialized Mammals*. Kon. Vlaamse Acad., Wentensch, Lett. Schone Kunsten Belg. pt. I:97–116.

Parker, R. B., and H. Toots, 1980. "Trace elements in bones as paleobiological indicators." In A. K. Behrensmeyer and A. Hill, eds., *Fossils in the Making*. University of Chicago Press, Chicago.

Parmalee, P. W., R. D. Oesch, and J. E. Guilday, 1969. "Pleistocene and recent vertebrate faunas from Crankshaft Cave, Missouri." *Reports of Investigations*, no. 14, Illinois State Museum, Springfield, Ill.

Payne, S., 1972. "On the interpretation of bone samples from archaeological sites." In E. Higgs, ed. *Economic Prehistory*. Cambridge University Press, London.

Payne, J. S., 1965. "A summer carrion study of the baby pig *Sus scrofa* Linnaeus." *Ecol.* 46:562–602.

Pennycuick, C., 1976. "Breeding of the lappet-faced and white-headed vultures. (*Torgos tracheliotus* Forster and *Trigonoceps occipitalis* Burchell) on the Serengeti Plains, Tanzania." *E. Afr. Wildl. J.* 14(1)67–84.

Phillips-Conroy, J. E., 1978. "Dental Variability of Ethiopian Baboon Populations: Anubis, Hamadryas, and Hybrid Baboons of the Awash National Park, Ethiopia." Unpublished Ph.D. thesis, New York University.

Pienaar, U. de V., 1969. "Predator-prey relationships amongst the larger mammals of the Kruger National Park." *Koedoe* 12:108–176.

Pilbeam, D. R., A. K. Behrensmeyer, J. C. Barry, and S. M. Ibrahim Shah, 1979. "Miocene sediments and faunas of Pakistan." *Postilla* 179:1–45.

Ramsay, J. G., 1967. *Folding and Fracturing of Rocks*. McGraw-Hill, New York.

Raup, D., and S. M. Stanley, 1971. *Principles of Paleontology*. W. H. Freeman, San Francisco.

Rolfe, W. D. I., and D. W. Brett, 1969. "Fossilization processes." In G. G. Eglinton and M. T. J. Murphy, eds., *Organic Geochemistry: Methods and Results*. Springer-Verlag, Berlin.

Roper, M., 1969. "A survey of the evidence for intrahuman killing in the Pleistocene." *Curr. Anthrop.* 10(4) pt.II:427–459.

Rosenthal, H. L., 1963. "Uptake, turnover and transport of bone-seeking elements in fishes." *N.Y. Acad. Sci. Annals* 109:278–293.

Rudnai, J., 1973. *The Social Life of the Lion*. Washington Square East, Wallingford, Pa.

Sadek-Kooros, H., 1972. "Primitive bone fracturing; a method of research." *Am. Antiq.* 37:369–382.

Salt, G. W., 1967. "Predation in an experimental protozoan population (*Woodruffia-Paramecium*)." *Ecol. Monogr.* 37:113–114.

Saunders, J. J., 1977. "Late Pleistocene vertebrates of the western Ozark highland, Missouri." *Reports of Investigations* no. 33, Illinois State Museum, Springfield, Ill.

Schaller, G., 1972. *The Serengeti Lion: A Study of Predator-Prey Relations*. University of Chicago Press, Chicago.

Selley, R. C., 1970. *Ancient Sedimentary Environments: A Brief Survey*. Cornell University Press, Ithaca, N.Y.

Shipman, P., in press. "Applications of scanning electron microscopy to taphonomic problems." *N.Y. Acad. Sci. Ann.*

Shipman, P., 1979. "Microscopic effects of known taphonomic events on bone and teeth." Paper presented to American Association of Physical Anthropologists, San Francisco.

Shipman, P., 1977. "Paleoecology, taphonomic history and population dynamics of the vertebrate assemblage from the middle Miocene of Fort Ternan, Kenya." Unpublished Ph.D. thesis, New York University.

Shipman, P., 1975. "Implications of drought for vertebrate fossil assemblages." *Nature* 257:667–668.

Shipman, P., and J. E. Phillips-Conroy, 1977. "Hominid tool-making versus carnivore scavenging." *Am. J. Phys. Anthrop.* 46:77–86.

Shipman, P., and A. Walker, 1980. "Bone-collecting in harvesting ants." *Paleobiol.* 6(4):496–502.

Shipman, P., W. Bosler, and K. L. Davis, 1981. "Butchering of giant geladas at an Acheulian site." *Curr. Anthrop.* 22(3):1–10.

Shipman, P., A. Walker, J. A. Van Couvering, P. J. Hooker, and J. A. Miller, 1981. "The Fort Ternan hominoid site, Kenya: geology, age, taphonomy and paleoecology." *J. Hum. Evol.* 10:1–28.

Shotwell, J. A., 1958. "Inter-community relations in Hemphillian (Mid-Pliocene) mammals." *Ecol.* 39(2):271–282.

Shotwell, J. A., 1955. "An approach to the paleoecology of mammals." *Ecol.* 36(2)327–337.

Simpson, G. G., 1970. "Uniformitarianism: An inquiry into principle, theory and method in geohistory and biohistory." In M. K. Hecht and W. C. Steere, eds., *Essays in Evolution and Genetics*. Appleton-Century Crofts, New York.

Simpson, G. G., 1963. "The meaning of taxonomic statements." In S. L. Washburn, ed., *Classification and Human Evolution*. Wenner-Gren Foundation for Anthropological Research, New York.

Simpson, G. G., 1960. "Notes on the measurement of faunal resemblance." *Am. J. Sci.* 258A:300–311.

Simpson, G. G., 1947. "Holoarctic mammalian faunas and continental relationships during the Cenozoic." *Geol. Soc. Am. Bull.* 58:613–688.

Simpson, G. G., 1946. "Bones in the brewery." *Nat. Hist.* June:252–259.

Simpson, G. G., A. Roe, and R. C. Lewontin, 1960. *Quantitative Zoology*. Harcourt, Brace, New York.

Skinner, J. D., S. Davis, and G. Ilani, 1979. "Bone collecting by striped hyaenas (*Hyaena hyaena*) in Israel." Paper presented to South African Society for Quaternary Archaeologists meeting.

Spinage, C. A., 1973. "A review of the age determination of mammals by means of teeth, with especial reference to Africa." *E. Afr. Wildl. J.* 11:165–187.

Stewart, J. M., 1977. "Frozen mammoths from Siberia bring the Ice Ages to vivid life." *Smithsonian Magazine*, Dec.:60–69.

Stewart, T. D., 1979. *Essentials of Forensic Anthropology*. Charles C. Thomas, Springfield, Ill.

Sutcliffe, A., 1970. "Spotted hyaena: crusher, gnawer, digestor and collector of bones." *Nature* 227:1110–1113.

Tappen, N. C., 1976. "Advanced weathering cracks as an improvement on split-line preparations for analysis of structural orientation in compact bone." *Am. J. Phys. Anthrop.* 44(2):375–380.

Toots, H., 1965a. "Orientation and distribution of fossils as environmental indicators." *Wyo. Geol. Assn. Guidebook*, 19th field conference: 219–229.

Toots, H., 1965b. "Sequence of disarticulation in mammalian skeletons." *Contrib. Geol.* 4(1):37–38.

Toots, H., and M. R. Voorhies, 1965. "Strontium in fossil bones and the reconstruction of food chains." *Science* 149:849–855.

Van Couvering, J. A. H., and J. A. Van Couvering, 1976. "Early Miocene mammal fossils from East Africa: aspects of geology, faunistics and paleoecology." In G. Ll. Isaac and E. R. McCown, eds., *Human Origins: Louis Leakey and the East African Evidence*. W. A. Benjamin, Menlo Park, Calif.

Van Lawick-Goodall, J., and H. Van Lawick, 1970. *Innocent Killers*. Collins, London.

Van Valen, L., 1964a. "Selection in natural populations: *Merychippus primus*, a fossil horse." *Nature* 197:1181–1183.

Van Valen, L., 1964b. "Relative abundance of species in some fossil mammal faunas." *Am. Nat.* 98:108–116.

Voorhies, M. R., 1969. "Taphonomy and population dynamics of the early Pliocene vertebrate fauna, Knox County, Nebraska." *Contrib. Geol. Spec. Paper*, no. 1.

Vrba, E. S., 1980. "The significance of bovid remains as indicators of environment and predation patterns." In A. K. Behrensmeyer and A. Hill, eds., *Fossils in the Making*. University of Chicago Press, Chicago.

Vrba, E. S., 1975. "Some evidence of chronology and paleoecology of Sterkfontein, Swartkrans and Kromdraai from the fossil Bovidae." *Nature* 254:301–304.

Walker, A., in press. "Dietary hypotheses and human evolution." *Trans. Roy. Soc. London*.

Walker, A., 1980. "Functional anatomy and taphonomy." In A. K. Behrensmeyer and A. Hill, eds., *Fossils in the Making: Vertebrate Taphonomy and Paleoecology*. University of Chicago Press, Chicago.

Walker, A., 1979. "S.E.M. analysis of microwear and its correlation with dietary patterns." Paper presented to American Association of Physical Anthropologists, San Francisco.

Walker, A., 1976a. "The Hunter Hunted." *Nat. Hist.* May:76–81.

Walker, A., 1976b. "Remains attributable to *Australopithecus* in the East Rudolf succession." In Y. Coppens, F. C. Howell, G. Ll. Isaac and R. E. F. Leakey, eds., *Earliest Man and Environments in the Lake Rudolf Basin: Stratigraphy, Paleoecology and Evolution.* University of Chicago Press, Chicago.

Walker, A., H. N. Hoeck, and L. Perez, 1978. "Microwear on mammalian teeth as an indicator of diet." *Science* 201:908–910.

Walker, P., and J. C. Long, 1977. "An experimental study of the morphological characteristics of tool marks." *Am. Antiq.* 42(4):605–618.

Western, D., 1979. "Size, life history and ecology in mammals." *Afr. J. Ecol.* 17(4):185–204.

Western, D., 1973. "The structure, dynamics and changes of the Amboseli ecosystem." Unpublished Ph.D. dissertation, University of Nairobi.

White, T. E., 1955. "Observations on the butchering technique of some aboriginal peoples, nos. 7, 8, 9." *Am. Antiq.* 21:170–178.

White, T. E., 1954. "Observations on the butchering techniques of some aboriginal peoples, no. 3, 4, 5, 6." *Am. Antiq.* 19:254–264.

White, T. E., 1953. "Observations on the butchering techniques of some aboriginal peoples, no. 2." *Am. Antiq.* 19:160–164.

White, T. E., 1952. "Observations on the butchering techniques of some aboriginal peoples, no. 1." *Am. Antiq.* 17:337–338.

Wolff, R. G., 1975. "Sampling and sample size in ecological analysis of fossil mammals." *Paleobiol.* 1(2):195–204.

Wood, B. A., 1976. "Remains attributable to *Homo* in the East Rudolf succession." In Y. Coppens, F. C. Howell, G. Ll. Isaac and R. E. F. Leakey, eds., *Earliest Man and Environments in the Lake Rudolf Basin: Stratigraphy, Paleoecology and Evolution.* University of Chicago Press, Chicago.

Yellen, J. E., 1977. "Cultural patterning in faunal remains: evidence from the !Kung Bushmen." In D. Ingersoll, J. E. Yellen, and W. MacDonald, eds., *Experimental Archeology.* Columbia University Press, New York.

Ziegler, A. C., 1973. "Inference from prehistoric faunal remains." Addison-Wesley publication no. 43.

Zumberge, J. H., and C. A. Nelson, 1972. *Elements of Geology.* John Wiley and Sons, New York.

Index

Abrasion, 41, 100, 113–115; categories of, 114
Aeolian transport, 41, 70, 100, 113
Afrocricetodon songhori, 126–127
Age: as cause of death, 20; structure, 156–169; determination, 160–162
Agent: of concentration, 65, 131, 145, 146; of breakage, 104–108; defined, 199
Allochthonous, defined, 199
Alluvium, 50; defined, 199
Amboseli National Park, 7, 115–119
Assemblages, 5, 62; taphonomic history of, 99–121; archaeological, 123–125, 192; comparison, 129–130, 137–141, 146–155; primary and secondary, 155; defined, 199
Australopithecines: bones of, 17, 179–181; in South African caves, 46, 58–60, 104; reconstructions of, 193, 194
Autochthonous, 31; defined, 199

Biocoenose, defined, 199
Bioturbation, 56, 69–70; defined, 199
Bone: mechanical properties of, 21–22, 104–108; size, 22–23, 30, 106; composition, 23–26, 30; density, 23, 25, 37–38; shape, 26–28, 37, 106; volume, 22, 37

Bones, indeterminate, 128–131
Boney Spring, Missouri, 57, 58, 70, 77
Breakage of bones: prefossilization, 100, 104–108, 172–179; types, 105, 106–107; postfossilization, 172–179
Breccia-splitting machine, 185
Bukwa, Uganda, 27–29, 56, 79, 95–96
Burning of bone, 177–179
Butchering by hominids, 124; in Wyoming, 86–88; at Olorgesailie, 107–108

Caliche deposits, 61
Carnivores, 18–19, 22–23, 25, 39–41, 91, 95, 99–100, 103, 104; scat of, 39–40; marks left by, 108–113, 192; strontium content of bones, 119–120; assemblages collected by, 132–137, 155–156
Caves, developmental sequence of, 55, 57–63
Channel deposits, 50–52, 56, 62–63; at Fort Ternan, 82–86; defined, 199
Chewing marks, 100, 108–113. *See also* Tooth marks
Chopping marks, 108–110; defined, 199
Collecting techniques, 184–189
Collector bias, 101, 128, 189

Communities, animal, 145–146, 191–196; comparisons among, 146–152; defined, 199
Continental margins, 53–55
Correlation coefficient, 137–141; defined, 199
Cracking, mosaic, 177–179; defined, 200
Crankshaft Pit, Missouri, 60, 61
Critical shear stress, 33–41; formula for, 35; defined, 200
Crown height measurement, 161–162, 164–165
CSI (Corrected number of Specimens per Individual), 143–145; defined, 200
Current velocity, 30–31, 39
Cutting marks, 100, 108–113, 124; defined, 200

Death, modes of, 17–21, 196. *See also* Mortality
Deformation, plastic, 172, 179–181; defined, 200
Delta, 50–55, 62; defined, 200
Density, bone, 23; and hydraulic behavior, 36–38; defined, 200
Deserts, 55, 60–63
Diagenetic factors, 129, 171–172
Diameter, nominal, 33–41; formula, 34; defined, 200
Dip of fossils, 66, 69, 76–78; at Fort Ternan, 78, 85–86, 88; defined, 200
Disease, 15, 19–20
Distribution, *see* Spatial distribution
DSE (Distinctive Skeletal Element), 126–128; defined, 200

East Turkana, Kenya, 17, 55, 67, 107–108; modern bones at, 115–117
Efremov, J. A., 5–6
Environments, geologic, 48–62; aqueous and semiaqueous, 55–56; lacustrine, 55–56, 62–63; sedimentary, 62–63
Erosion, 181–184
Evidence versus theory, 12–16
Excavation techniques, 184–188

Fauna, 6; comparisons of, 148–152; representation of, 186–189; defined, 200
Faunal analysis, 123–170
Flumes, artificial, 31, 71
Fluvial environments, 42, 49–52, 62–63; characteristics of deposits, 52; defined, 200
Fort Ternan, Kenya, 27–29, 42, 47, 52, 66, 68; distribution of fossils, 79–87; dip of fossils, 85–86, 88; tooth marks on fossils, 113; weathering of fossils, 115, 117; indeterminate fragments, 130–131; faunal representation, 186–187
Fossil: defined, 1; compared with living animals, 2–4
Fossilization, *see* Impregnation; Permineralization
Fracture: spiral, 104–106, 119, 174–175; columnar, 119, 173–174; perpendicular smooth, 174–177; defined, 200–201
FRI (Faunal Resemblance Index), 148–149; defined, 201

Habitat preference, 152–156, 192
Harvester ants, 22, 95, 132–133
Historical sciences, 9–12
Hominids: bones of, 17, 104; use of bones by, 41; in caves, 58–60; influence on spatial distribution, 86–95; bone breakage by, 104–108; diet, 119–120, 123–125
Hydraulic behavior: of bones, 28–41; and transport of assemblages, 85, 88–90, 113; defined, 201
Hydrodynamic sorting, 30–33, 35, 62, 67, 131, 155; defined, 201

Identification: by skeletal element, 126–128; by taxon, 126–128
Igneous rocks, 45, 47–48; defined, 201
Impact law, 33
Impregnation, 172; defined, 201
Information loss, 12–15
Insect holes, 111, 112

Klasies River Mouth, South Africa, 166–169
Kolmogorov-Smirnov test, 166; defined, 201
K-selection, 159; defined, 201

La Brea tar pits, California, 20, 43, 56
Lacustrine environments, 55–56, 62–63; defined, 201
Lag deposit, 31–33, 50, 56, 61–62, 104; defined, 201
Langebaanweg, South Africa, 55
Life table, 157–159; defined, 201

Metamorphic rocks, 45, 48; defined, 201
Microvertebrates, 13–16, 39–41, 95, 132–133, 192–193
MNI (Minimum Number of Individuals), 142–145, 155–156, 192; defined, 201
Mortality: attritional, 18, 157–169, 193; catastrophic, 18, 157–169, 193; defined, 201–202

Napak, Uganda, 52–53, 77
Nelson Bay, South Africa, 166–169
Neotaphonomy, 7; defined, 202
NSI (Number of Specimens per Individual), 143–144; defined, 202

Omo, Ethiopia, 56, 70, 120
Olduvai, Tanzania, 40, 55, 70; stone circles at, 90–95; bone breakage at, 107–108; indeterminate bones at, 130–131
Olorgesailie, Kenya, 55, 87, 107–108
Orientation, 69–76; patterns of, 73–76; defined, 202
Overbank deposit, 50–52; defined, 202
Oxbow lake, 50–52, 62; defined, 202

Paleoecology: defined, 3, 202; as historical science, 9–12; reconstructing, 192–196
Paleomagnetic dating, 47–48; defined, 202
Paleotaphonomy, 6; defined, 202

Pearson's r, 137–141; defined, 202
Peat bogs, 1, 55–56, 62–63
Percentage representation, 133–134; defined, 202
Permafrost, 61–62
Permineralization, 53, 172; defined, 202
Phreatic zone, 58–59; defined, 202
Point bar, 50–51; defined, 202
Population dynamics, 156–169; see also Life table; MNI; Mortality; Survivorship curve
Porcupines as bone collectors, 12, 23, 95–96, 133–137, 155–156; chewing marks of, 111
Postdepositional events, 17, 171–196; dipping, 69; defined, 202
Postfossilization events, 17, 171–196; damage, 56; deformation, 179–182; defined, 202
Postmortem period, 17, 21; events, 41, 60; crushing, 56, 172, 179–181; defined, 202
Predation, 18–19, 192; curve, 166–169, 203. See also Carnivores; Raptor pellets
Preservation: differential, 17, 41–43, 62–63, 99–104, 142; potential, 21, 25, 42–43, 126–128, 203; state of, 99, 120–121, 130
Prodelta, 53–55; defined, 203
Puncture, 99, 108–110; defined, 203
Pyroclastics, 47

Radiometric dating, 47–48
RAI (Relative Abundance Index), 149; defined, 203
Raptor pellets, 15–16, 22, 39–41, 95, 132–133, 192–193
Representation: of bones, 100–104; skeletal, 131, 141, 155
Retrodiction, 10–11
Rose diagrams, 71–77
r-selection, 159; defined, 203
Rusinga, Kenya, 52, 70

Sampling, X-type, 152–156; defined, 203

Index / 221

Sampling, Y-type, 152–156; defined, 203
SA/V ratio (Surface Area to Volume ratio), 26, 30–31, 128; defined, 203
Scanning electron microscopy: of bone, 108–114; of teeth, 193, 195–196
Scavengers, 22, 95, 100, 103, 111, 129
S/C ratio (Spongy to Compact bone ratio), 25, 34–35, 101–102, 128; defined, 203
Sedimentary: rocks, 45–47; particles, 45–47, 50; environments, 48–63; defined, 203
Senility, 20
Settling velocity, 33–35, 39; formula, 33; defined, 203
SI (Shape Index) of bones, 26–29; defined, 203
Sieving techniques, 185–188
Sinkholes, 55, 57–63
Siwaliks, Pakistan, 50
Skeletal elements, 131–141; primate, 102; bovid, 103. *See also* Identification; Representation
Slicing mark, 108–113; defined, 204
Spatial distribution, 65–98; data, 65; horizontal, 66–67; vertical, 68–69; orientation, 69–76; result of geologic events, 79–85; result of hominid activities, 86–95; result of animal activities, 95–97; result of temporal events, 97. *See also* Dip of fossils
Species: abundance, 142–145, 192; diversity, 146–149, 187–188; plasticity, 195
Springs, 55, 56, 76
Square frequency, 67; defined, 204
Starvation, 21
Steppes, 61–62

Strike, 78; defined, 204
Strontium analysis, 100, 119–120, 193
Survivorship curve, 159–160; defined, 204
Swamps, 55–56, 62–63

Taphocoenose, defined, 6, 204
Taphonomic history, 12–14, 42–43, 99–121, 192; defined, 204
Taphonomy, defined, 6, 204; as historical science, 9–12
Tar pits, 43, 55–56, 62–63
Taxon, identification of, 126–128
Tectonic areas, 48, 49–52
Teeth: preservation potential of, 128; and diet, 195–196
Thanatocoenose, defined, 6, 204
Threshold drag, *see* Critical shear stress
Tooth marks, 99–100, 108–113, 132–137, 155, 192–193; defined, 204
Trampling, 95, 104, 119, 173–174, 179
Transportation of bones, 23
Trophic replacement, 150–152; defined, 204
Tuff, 47, 52; defined, 204
T/V ratio (Teeth to Vertebrae ratio), 62; defined, 204

Uniformitarianism, 11–12

Vadose zone, 58–59; defined, 205
Verdigre Quarry, Nebraska, 69, 164–169
Volcanic fields, 52–53
Voorhies Groups, 31–33, 38–39, 131; defined, 205

Weathering, 41, 100, 115–119, 174, 177–179; stages of, 116, 117